HOW TO FIND
INFORMATION
in science and technology

Second edition

HOW TO FIND
INFORMATION
in science and technology

Second edition

Jill Lambert
Information Services, Birmingham Polytechnic

Peter A. Lambert
Department of Pharmaceutical Sciences, Aston University

LIBRARY ASSOCIATION PUBLISHING
LONDON
A Clive Bingley Book

Published by
Library Association Publishing Ltd
7 Ridgmount Street
London WC1E 7AE

First published 1986
This second edition 1991

British Library Cataloguing in Publication Data

Lambert, Jill
  How to find information in science and technology.
  – 2nd ed.
  I. Title II. Lambert, Peter
  507

ISBN 0-85157-469-6

Typeset in 10/12pt Palacio by Library Association Publishing Ltd
Printed and made in Great Britain by Billing and Sons Ltd, Worcester

For Cathy and Ian

Contents

Acknowledgements   viii

Illustration acknowledgements   ix

1   How information is communicated   1

2   Beginning a search   11

3   Using abstracts and indexes   31

4   Computerized information searching   51

5   Obtaining and organizing information   73

6   Keeping up to date   89

7   Future developments in scientific and technical information   101

Index   105

Acknowledgements

We would like to thank Graham Smith of the Department of Pharmaceutical Sciences, Aston University, for preparing the figures and cartoons. The idea for Figure 34 came from *Personal bibliographic indexes and their computerisation* by R. Heeks, published in 1986 by Taylor Graham. Our thanks are due to all the organizations who generously allowed us to reproduce examples of their publications.

Jill Lambert
Peter A. Lambert

Illustration acknowledgements

Figure
2 Reproduced from the *McGraw-Hill encyclopedia of science and technology*, McGraw-Hill Inc., 1987, with the permission of McGraw-Hill Inc., New York.
3 Reproduced from Mazda, F. F., *Electronics engineer's reference book*, 6th ed., Butterworths, 1989, with the permission of Butterworth-Heinemann, Guildford.
8 Reproduced from *Whitaker's books in print*, Whitaker, 1990, with the permission of J. Whitaker & Sons Ltd, London.
9 Reproduced from *Subject guide to books in print*, Bowker, 1989–90, with the permission of Bowker-Saur, London.
10 Reproduced from *British national bibliography*, British Library, 1988, with the permission of the British Library National Bibliographic Service, London.
11 *Cumulative book index*, 1988, Volume 1, A–K. Copyright © 1988 by the H. W. Wilson Company. Material reproduced by permission of the publisher.
12 Reproduced from the *Index to scientific reviews 1989*, with permission from and copyright owned by the Institute for Scientific Information[R], Philadelphia, PA.
13 Reproduced from *Index to theses*, Aslib, 1988, with the permission of Aslib, London.
14 Reproduced from *Dissertation abstracts international*, University Microfilms International, 1990, with the permission of University Microfilms Inc., Ann Arbor, MI.
15 Reproduced from: *Index medicus*, National Library of Medicine, 1990, with the permission of the National Library of Medicine, National Institutes of Health, Bethesda, MD.; *Chemical abstracts*, Chemical Abstracts Service, 1990, with the permission of Chemical Abstracts

1

How information is communicated

Our purpose in writing this book is to help the practising scientist or technologist to find information. Before we begin to outline the methods and techniques used to find information, though, we must first take one step back and consider how scientific, technical and medical information is communicated. This information, generated by a variety of organizations including industrial and commercial enterprises and academic and government institutions, reaches the outside world through several different channels. Searching for information is easier once you know what these channels are and how they interlock to form a communication system.

People

The first obvious but very important way of spreading information is by word of mouth. People talk to friends and colleagues, passing on news about their own work and often including pieces of information they have read elsewhere. Information is also spread in this way in a more organized fashion by people talking at seminars, conferences and so on.

Picking up information from friends and going to meetings is very popular, particularly in engineering. The reasons are simple enough – it is far more convenient to ask someone working in the same field than it is to search through a mass of literature, and easier to learn about new work at a meeting than to read about it in print. The snag with transferring information in this way is that it can only reach a very limited number of people. If you need to know something and your contacts cannot help, you have a problem. That is why other printed methods of communicating information, which can reach a large audience, are also needed.

Journals

New research which does not have to be kept secret for commercial or defence reasons is nearly always published in the form of articles or papers in journals. There are a large number of journals publishing research in this way, and they range in scope from very general ones covering the whole field of science to the very specific, covering small subject specialisms. Some are published by learned societies and professional institutions such as the Royal Society of Chemistry and the Institution of Mechanical Engineers, while others are the product of commercial publishers such as Blackwells Scientific, Elsevier Science, Pergamon, Wiley-Interscience and Academic Press. Before a paper is considered worthy of publishing it has to be refereed, that is, evaluated by fellow-workers in the subject field. Refereeing practices vary between journals but generally manuscripts are evaluated by two independent assessors who are looking for originality, validity and quality.

Most of the space in these journals is taken up by full papers having a well-defined format, with an introduction in which previous work is outlined and the objective of the research project indicated. This is followed by a section outlining the experimental work conducted and any special methods used, the results, and the conclusions discussing their significance and value. Where the findings are more limited, these are sometimes published as short papers, usually four to five pages in length.

The refereeing procedures plus the time required for editing, printing and binding mean that most papers are published approximately six to nine months after their submission. Where authors want research made public very quickly, perhaps because they believe it to be particularly significant or because there are rival teams working in the same field, they can attempt to publish the work in the form of a preliminary communication or letter. Such communications tend to have fairly brief introductory and experimental sections, concentrating on the results and conclusions which can be drawn from the work. In order to reduce delays in publication, the refereeing procedure is usually abbreviated for such communications. It is assumed, however, that these will be followed later by full papers, although in practice this is by no means always the case.

Many journals publish a mixture of all three types of paper: full, short and preliminary. With the rise in popularity of preliminary communications, new journals consisting solely of this type of paper have been established. To minimize the length of time between submission and publication of a paper, many of these journals use camera-ready copy, that is, the typescript is prepared by the author, on whom the onus is placed to produce a perfect copy.

Not all journals limit themselves to publishing research papers in the form outlined. Some journals concentrate on picking out the more interesting new developments and presenting these in a readable manner. Two of the most familiar are *New scientist* and *Scientific American*, but there are many more specialized ones, particularly within engineering. Apart from the fact that they make interesting reading, these journals can be useful because new products are widely advertised in them.

Theses

Much of the research published in journals is carried out by students working for higher degrees. This work has to be written up in the form of a thesis or dissertation, so that it can be assessed by external examiners. Theses can be good sources of information because the first chapter is always a state-of-the-art review of the subject, backed up by a comprehensive list of references to previous work.

Conference proceedings

Very often the research reported at conferences is later published as proceedings: sometimes in the form of a book, sometimes as separate papers in a journal. Conference proceedings can be a useful way of finding out about projects still in progress which have not been written up in complete form as papers in journals. Conferences tend to focus on expanding subject fields or themes which are becoming increasingly important, so the published proceedings often reflect the future direction of a speciality.

One note of caution here – published proceedings, at least those published in book form, are not refereed as are papers published in journals. The data may not therefore have been subjected to the same critical scrutiny.

Reports

A significant proportion of applied research and development is first written up in the form of reports. Some reports are made openly available to anyone interested but where the work is of commercial or military value they are restricted to a strictly specified limited number of people.

When reading reports it is worth remembering, as with conference proceedings, that the information has not been externally evaluated or refereed. Much of the information which first appears in reports is in fact later published in journals, but in condensed versions. If you have a paper in which reference is made to an earlier report on the topic, it could be worth getting hold of this because of the extra detail included.

Patents

Industrial research and development which has some commercial value is normally published as patents. These are documents granted by governments to the owners of an invention, allowing them a monopoly for a limited period of time in order to exploit the invention. In return for this monopoly, the owners have to disclose, that is, make public, all the information available about the invention. Obtaining a patent is quite a lengthy and complicated affair since the invention has to be examined by a government's patent office. In the UK the first information about a patent application is generally published about 18 months after submission. The patent application is then formally examined and if accepted is republished as a complete specification two years later.

Despite the large numbers of patents in existence – approximately 2,000,000 British patents were published between 1917 and 1990, for instance – it must be acknowledged that they are not especially popular information sources for scientists or technologists. Problems are experienced with the legal style, and there are complaints about inadequate descriptions and implausible claims. There does seem to be a general belief that the information will later appear as papers in journals or conference proceedings. Certainly some information is later incorporated into promotional or trade literature, but that apart, very little is republished elsewhere. Although there may be difficulties in extracting the information contained in patents, to ignore their existence completely is unwise.

Trade literature

Trade literature – advertisements, catalogues and company magazines – is a well-used source of information, particularly in engineering and construction. It contains the kind of practical information not published in more conventional literature, and also has the plus point of (normally) being well presented and illustrated.

Standards

Most of the publications we have outlined so far are used to spread information about new research and development. Standards are different: they do not communicate new facts but specify acceptable dimensions in a product or set acceptable levels of quality or codify good existing practices. The most familiar, at least in the UK, are those produced by the British Standards Institution, but there are many other organizations, worldwide, which are also involved in formulating standards.

Books

The most familiar of all the printed methods of spreading information is, of course, books. Books do not report directly on new research or development: their publication schedule would be too slow and the information content of the average research paper would be too specialized to be of interest to sufficient readers to make a book economically viable.

What the author of a book does is to repackage and evaluate information which has already appeared in a lot of different publications, which makes the book very useful when you need an introduction to an unfamiliar topic or to check a fact.

Reviews

Another way of drawing together information appearing in journals, conference proceedings and so on is by means of a review: a critical evaluation of the developments which have taken place in a speciality. Reviews are very highly regarded as information sources by scientists and technologists because they are written by acknowledged experts, and are well provided with references to further readings.

Reviews can be found singly in all sorts of publications – journals, conference proceedings and so on – and also collected together in separate review serials. These usually have easily recognizable titles such as *Annual review in* ..., *Advances in* ... and *Progress in* ....

Abstracts, indexes and databases

All these different channels of communication are potential sources of information. The problem is how to find what you want among the mass of irrelevant items. One method of looking for information on a topic would be to look at as many relevant journals as you could find; another method might be to browse through any conference proceedings available. Searching in this way is not very systematic though; it can take a long time and the result will depend very much on what material is actually on the shelves in your library or information centre.

A much more efficient, effective way is to use abstracts or indexes. These are publications listing papers appearing in many different journals. Many also include conference proceedings, and some include reports and patents. There are many of these abstracts and indexes in science and technology – virtually every subject field is covered by one or more of them – and they are produced in the form of journals, coming out at weekly, fortnightly or longer intervals throughout the year. Most are now also produced in electronic form, searchable by computers. In this form they are usually known as databases or databanks. As Figure 1 shows, they play a major role in the network by which scientific and technical information is communicated.

References

Before outlining the practical methods you can use to find information, it is worth briefly mentioning how work is cited, that is, referred to or acknowledged in publications.

There are two different ways of citing references.

1   The numeric system, in which each publication referred to is given a separate number; for example, 'Further work (9) has shown' or 'Further work by Jones (9) has shown'. The numbers can also be given as superscripts. The references are then listed in full in numerical order at the end of the publication.

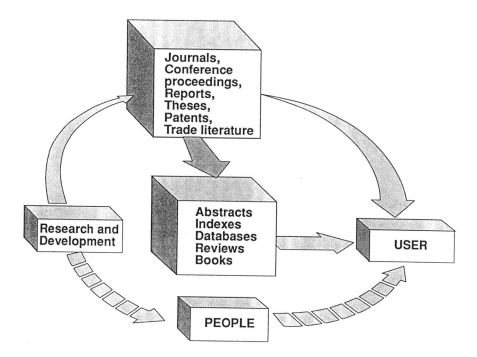

Fig. 1 Communication of scientific and technical information

2   The Harvard system, in which each time a publication is referred to the author's name and year of publication are given; for example, 'Further work (Jones 1990) has shown' or 'Further work by Jones (1990) has shown'. If there are two authors, both names are included. Where there are more than two authors, only the first name is quoted, followed by *'et al.'*. The references are listed in full in alphabetical order of author's names at the end of the publication.

There is no one standard way of setting out a reference: practices vary between publishers. This does not matter too much as long as sufficient information is included about each publication to allow it to be identified uniquely and traced.

Books

References should include the author, title, publisher, year of publication and preferably the town or city of publication, as below:

BRAITHWAITE, N. and WEAVER, G., *Electronic materials*, Sevenoaks, Butterworth, 1990.

6

Where an author has contributed to a book which has been edited by a different author, details of both contributor and editor are included; for example:

SHARKEY, J., 'Building blocks of a fourth-generation system', in HOLLOWAY, S. (ed.), *Fourth generation systems*, London, Chapman and Hall, 1990, pp. 57–73.

Reports and theses are also referred to in a similar way. Usually reports are given code letters and numbers – such as ST1/DOC/10/246 – which should be included in the reference. References to theses should state the type of degree, such as MSc or PhD, and the institution where the work was carried out.

Journals
Because there are so many journals, each issue of which can contain a large number of papers, it is necessary to have a system by which each paper can be referred to precisely. This means including not only the author and title of the paper but also the name of the journal and details of which part or issue it appeared in, as in the reference below:

WATSON, J. D. and CRICK, F. H. C., 'A structure for deoxyribose nucleic acid', *Nature*, **171** (4356), 1953, 737–8.

The volume number (the number given to all the separate issues published in any one year) usually appears in either italic or bold print. Titles of journals are printed in italic to distinguish them from the titles of papers. Journal titles are very frequently abbreviated, but unfortunately the recommended standard abbreviations are not always used.

If you are in doubt as to what an abbreviation stands for, it is better not to guess: a lot of time has been wasted chasing inaccurate references in this way. Abbreviations can be checked in several books, a recommended one being A. Alkire's *Periodical title abbreviations*, 7th edition, Detroit, Michigan, Gale Research, 1989.

Conference proceedings
References to papers published in conference proceedings tend to be fairly long. To avoid confusing two conferences with similar names, it is important to include the number of the conference (if there is one) and the name and place it was held at, as shown below:

WALTON, A. D., 'Jaguar's approach to the introduction of CIM', in *Computer integrated manufacturing: proceedings of the 4th CIM conference*, Madrid, 18–20 May 1988. Bedford, IFS Publications, 1988, pp. 85–98.

Patents

Every patent published is given a separate number, which should be included in the reference; for example:

ROLLS-ROYCE PLC, *Gas turbine engine tip clearance sensor systems*, GB patent application 2229004A, 1990.

Standards

Standards are also published bearing individual numbers and codes, which should also be included in the reference; for example:

BRITISH STANDARDS INSTITUTION, *Code of practice for fire precautions in the chemical and allied industries*, 1990, BS 5908.

2

Beginning a search

At the beginning of a new investigation, you will want to find as much information as possible about the topic. There are two reasons why it is best to do this in advance of practical or laboratory work. First, you are less likely to duplicate research which has already been carried out elsewhere. Secondly, any information you find about existing work can be used as the starting-point for the new investigation. It is always tempting, particularly for new researchers, to put off this type of search and to begin experimentation immediately, but doing this can produce some unpleasant surprises later on.

Before outlining how you can find this information, though, there are a couple of points worth mentioning. The first is that we have concentrated on outlining the general principles of information searching, illustrating these with examples from some major information sources. There are many more potential information sources which could not be included: your library or information centre will know of these sources and be able to give good advice on the best ones to use. Don't be reluctant to ask: the staff will be only too willing to help. The second point is that there is really no one single right way of searching for information, but a number of different methods, one or more of which will, it is hoped, produce results. What we aim to do is to point you in the right direction.

Using previously known work

The easiest way of starting is to make use of any previously known work on the topic. If you are joining an established research group, there is almost certain to be one person with some knowledge of the topic who will be able to pass on at least a few references to relevant papers. In the introductions to these papers, the authors will have outlined and given references to previous work in the subject field. At least some of these references will be relevant to the topic and provide you with a good starting-point for your research. It is worth scanning these references to see if you can spot a review article. Not all reviews are named as such, so you need to look fairly carefully. Since reviews tend to be much longer than research papers, one clue to their presence is their length, indicated by the page numbering. Any article which includes the word 'survey' in its title is also worth investigating. Following up these papers will produce a further set of references, some of which will probably be of use. Repeating this procedure once or twice should help you form an idea of the current stage of development in the subject.

It would be wrong, however, to give the impression that this type of search is sufficient by itself. Because the information has been collected in a fairly haphazard way, it is possible to miss important references. Also the search will not have picked up the very latest information: delays in publication plus the time required to prepare manuscripts mean that the most recent references included by authors will have appeared at least nine to twelve months previously.

What you need to do now is supplement this information by a more systematic search for books, reviews and papers. Knowing something about the field already will be a help in searching more selectively and efficiently.

Starting from scratch
Not everyone, though, is fortunate enough to begin an investigation with a handful of relevant references. If you do not find yourself in this position, do not be put off. If your knowledge of the topic is fairly limited, it is a good idea to spend a bit of time right at the start reading around the field and looking at the nomenclature used. One way of doing this is to look for a handbook or encyclopedia which covers the discipline. A large general one which should be easy to find is the *McGraw-Hill encyclopedia of science & technology* (Figure 2). Handbooks such as the *Electronics engineer's reference book*, shown in Figure 3, also tend to be particularly good sources of practical information. Don't assume that these kinds of publication are intended for the amateur or layman. They are generally written by experts in a subject for people with a scientific or technical education but new to the field. The articles themselves tend to be fairly terse and to the point and give references to some of the key texts in the field, for further reading.

Another good way for newcomers to a subject to orient themselves is to find a relevant guide to the published information. Libraries, learned societies and commercial publishers have all tried their hands at these guides, with the consequence that large numbers of them now exist. They range in size from a few word-processed pages to substantial books, depending on their origin and subject, but they all have the same basic purpose – to help you find out what has been published in the subject.

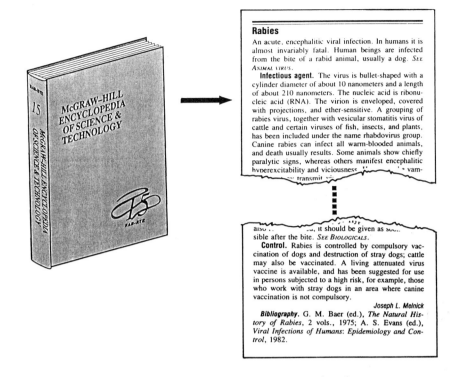

Rabies

An acute, encephalitic viral infection. In humans it is almost invariably fatal. Human beings are infected from the bite of a rabid animal, usually a dog. *SEE ANIMAL VIRUS.*

Infectious agent. The virus is bullet-shaped with a cylinder diameter of about 10 nanometers and a length of about 210 nanometers. The nucleic acid is ribonucleic acid (RNA). The virion is enveloped, covered with projections, and ether-sensitive. A grouping of rabies virus, together with vesicular stomatitis virus of cattle and certain viruses of fish, insects, and plants, has been included under the name rhabdovirus group. Canine rabies can infect all warm-blooded animals, and death usually results. Some animals show chiefly paralytic signs, whereas others manifest encephalitic hyperexcitability and viciousness   vam-   transmit

also   , it should be given as soon as possible after the bite. *SEE BIOLOGICALS.*

Control. Rabies is controlled by compulsory vaccination of dogs and destruction of stray dogs; cattle may also be vaccinated. A living attenuated virus vaccine is available, and has been suggested for use in persons subjected to a high risk, for example, those who work with stray dogs in an area where canine vaccination is not compulsory.

Joseph L. Melnick

Bibliography. G. M. Baer (ed.), *The Natural History of Rabies*, 2 vols., 1975; A. S. Evans (ed.), *Viral Infections of Humans: Epidemiology and Control*, 1982.

Fig. 2   *McGraw-Hill encyclopedia of science & technology*

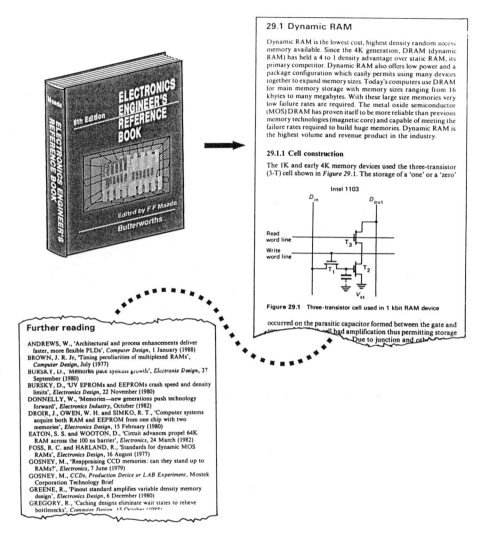

29.1 Dynamic RAM

Dynamic RAM is the lowest cost, highest density random access memory available. Since the 4K generation, DRAM (dynamic RAM) has held a 4 to 1 density advantage over static RAM, its primary competitor. Dynamic RAM also offers low power and a package configuration which easily permits using many devices together to expand memory sizes. Today's computers use DRAM for main memory storage with memory sizes ranging from 16 kbytes to many megabytes. With these large size memories very low failure rates are required. The metal oxide semiconductor (MOS) DRAM has proven itself to be more reliable than previous memory technologies (magnetic core) and capable of meeting the failure rates required to build huge memories. Dynamic RAM is the highest volume and revenue product in the industry.

29.1.1 Cell construction

The 1K and early 4K memory devices used the three-transistor (3-T) cell shown in *Figure 29*.1. The storage of a 'one' or a 'zero'

Figure 29.1 Three-transistor cell used in 1 kbit RAM device

occurred on the parasitic capacitor formed between the gate and ...cell had amplification thus permitting storage ...Due to junction and oth...

Further reading

ANDREWS, W., 'Architectural and process enhancements deliver faster, more flexible PLDs', *Computer Design*, 1 January (1988)
BROWN, J. R. Jr, 'Timing peculiarities of multiplexed RAMs', *Computer Design*, July (1977)
BURSKY, D., 'Memories pace systems growth', *Electronic Design*, 27 September (1980)
BURSKY, D., 'UV EPROMs and EEPROMs crash speed and density limits', *Electronics Design*, 22 November (1980)
DONNELLY, W., 'Memories—new generations push technology forward', *Electronics Industry*, October (1982)
DROIR, J., OWEN, W. H. and SIMKO, R. T., 'Computer systems acquire both RAM and EEPROM from one chip with two memories', *Electronics Design*, 15 February (1980)
EATON, S. S. and WOOTON, D., 'Circuit advances propel 64K RAM across the 100 ns barrier', *Electronics*, 24 March (1982)
FOSS, R. C. and HARLAND, R., 'Standards for dynamic MOS RAMs', *Electronics Design*, 16 August (1977)
GOSNEY, M., 'Reappraising CCD memories: can they stand up to RAMs?', *Electronics*, 7 June (1979)
GOSNEY, M., *CCDs, Production Device or LAB Experiment*, Mostek Corporation Technology Brief
GREENE, R., 'Pinout standard amplifies variable density memory design', *Electronics Design*, 6 December (1980)
GREGORY, R., 'Caching designs eliminate wait states to relieve bottlenecks', *Computer Design*, 15 October (1988)

Fig. 3 *Electronics engineer's reference book*

The benefit of using these guides is that they give a good grounding in the literature of a subject, which can then be supplemented with more recent material. Don't expect them to be either absolutely comprehensive or completely up to date, though. There is a good chance that as soon as a guide has been sent to the publishers, a flood of new books will appear, several journals will be launched and at least one important organization will change its name.

Library catalogues

Once you have discovered some relevant information about your topic – perhaps several journal references and a few books – you will want to read the originals as soon as possible. Although friends and colleagues will often be good sources of supply, you will need to obtain some items from a library, which inevitably means using the library's catalogue. Many people find catalogues very off-putting, which is a pity because the basic principles underlying their organization are both simple and effective. In the past, most catalogues, at least in the UK, took the form of 5 × 3 in. cards stored in catalogue drawers as in Figure 4a. Most library catalogues are now produced by computerized techniques and the cards have been replaced by computer-generated microfiches (Figure 4b). Online catalogues too have become commonplace in the last few years. In an online catalogue the records of a library's stock are recorded in an online database to which are connected many keyboards and VDUs (Figure 4c).

Name and title catalogues

Whatever the physical format, all catalogues have the same basic function – to let you find out if a book you want is in a library. To do this, a catalogue lists in alphabetical order the names of the authors and editors of books in stock along with the associated information about title, edition, publisher, year of publication and classification number. Most libraries also include the titles of books separately, so that if you only know the title of the book you can still find out if it is in stock. With common surnames such as Smith, Williams, Brown and so on it can be quicker to search for a book by its title than by its author.

Every book in a library has a classification number which represents the subject. Several different schemes have been devised for the classification of books. The commonest, and the one found in most public libraries, is the Dewey Decimal Classification. In this system general subject fields are represented by a three-figure number. Books on organic chemistry, for example, are given the number 547. Specialized subject fields are represented by longer numbers. Books on polymer science, a branch of organic chemistry, are given the number 547.84. Books are arranged on the shelves in classification number order: it is obviously necessary to find what classification number has been given to the book you want before looking for it on the shelves.

With an online catalogue a menu approach is used and the system will prompt you to enter the information needed. To find out if a certain book is in stock you have to type in the name or title on the keyboard. The computer then matches this with the records in the database, and a list of the most closely related items is displayed on the VDU. As each listed item is numbered, you need only key in the relevant number to obtain more

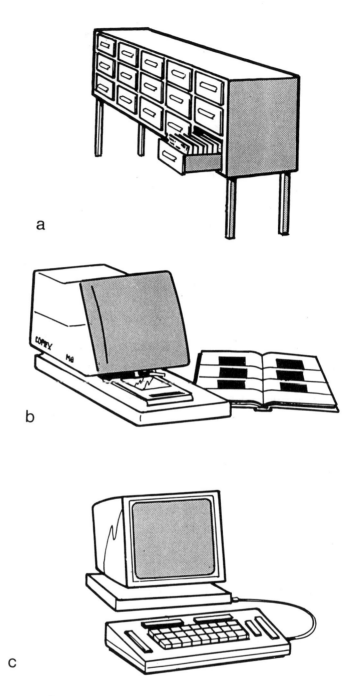

a

b

c

Fig. 4   Library catalogues

information about the book. In a card or microfiche catalogue, it is simply a question of looking for the right section or frame.

The layout of records in a catalogue can vary, but the example shown in Figure 5 is fairly typical. In addition to the classification number, there will usually be at least two other numbers. The International Standard Book Number (ISBN) is a unique identifying number given to each book at the time of publication. Accession or book numbers are numbers which libraries allocate to the books bought. Each separate copy bought will have a different number; the fact that there are two numbers in the example in Figure 5 means that the library has bought two copies. Some online catalogues will also tell you whether the book is actually in the library or whether it is currently on loan to someone.

There are two points worth noting about the way names and titles are filed in catalogues. The first, fairly obvious, one is that initial articles such as 'the', 'an' and 'a' are ignored for filing purposes. *The concise Oxford dictionary*, for example, will be listed under 'Concise. . .'. Perhaps less obvious but very important is the fact that words and phrases can be arranged by two different methods: letter by letter or word by word. With the letter-by-letter method, the spaces between the words are disregarded, whereas in the word-by-word method the spaces between words are significant.

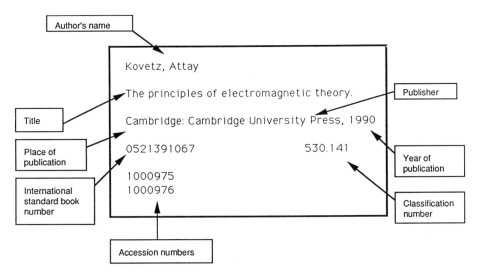

Fig. 5 Catalogue record

17

The different results produced by the two methods can be seen below.

letter by letter	word by word
database	data compression
database management systems	data transmission
databases	database
data compression	database management systems
data transmission	

Subject indexes

If you do not know the names of any authors or titles in the subject field, how can you find out if the library has any relevant books? The obvious thing to do is to walk round the shelves until you find the right place. If the library or information centre is well signposted, this approach should be fairly effective in finding general books in big subject fields such as electronic engineering or organic chemistry. It is not an effective way of finding books on specialized topics which, because there are so many, cannot be so clearly labelled. A better approach is to find out the appropriate classification number first from the subject index. Most libraries produce an index which links subjects to their classification numbers. An example of a typical index is shown in Figure 6.

A good subject index is a useful tool because it brings together all the different aspects of a subject; without using it, you run the risk of missing relevant books which approach the subject from a discipline different from your own.

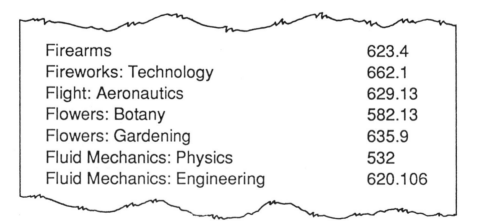

Firearms	623.4
Fireworks: Technology	662.1
Flight: Aeronautics	629.13
Flowers: Botany	582.13
Flowers: Gardening	635.9
Fluid Mechanics: Physics	532
Fluid Mechanics: Engineering	620.106

Fig. 6 Subject index

Subject catalogues

Tracing the books on the shelves should be a fairly quick procedure once you have found the appropriate class number or numbers. Normally you might expect to find at least one or two books on the topic, which could be sufficient if you are looking for background reading. It is impossible, though, to be certain, just by checking the shelves at the right classification number, that you have found all the relevant books: some will have been borrowed by other readers, and some will be missing because they are reserved, being repaired or awaiting reshelving. The only way to be sure of finding all the books on the topic is to check the subject catalogue. Although you might not have instant access to the books as they may be on loan, anything that looks interesting could be reserved so that it can be saved for you on return.

A subject catalogue, in microfiche or card format, records all books in a library in order by their classification numbers, so that all books on the same subject are together. The classification numbers are arranged in ascending numerical order; to see a list of books on a subject you need to find the right numbers on the fiches or cards. Online catalogues also list books by classification number, but in this case you are required to key in the appropriate number. Some online catalogues also include an extra facility – 'keyword' subject searching. With this facility, you can type in a 'keyword' (a word describing a subject), for example 'ultrasound'. The computer then matches this word with all the titles of books in the catalogue and displays on the VDU any books that include the keyword.

Journal catalogues

Many libraries and information centres include the names of all the journals subscribed to in the catalogue; most also produce a separate list. Sometimes this is on computer print-out, sometimes on microfiche and occasionally on cards, kept either in a drawer or file holder. Whatever the format, there will always be a record of each journal name and of the year and volume number when the library first began its subscription (Figure 7). Libraries do not always take a journal from the very first part or issue: very often a subscription may not start until after a journal has been in existence for some years. You do need to check the date of your journal reference with the year and volume number, to make sure it is in the library.

Lists of published books

It is unlikely that you would find every relevant book on your topic in a library or information centre. There will probably be books that you know of which have not been bought by the library. These can be borrowed for you by a procedure known as Inter-library loan, described in Chapter 5.

19

ENERGY POLICY	1980-, Volume 8- Bimonthly
ENERGY WORLD: BULLETIN OF THE INSTITUTE OF FUEL Previously JOURNAL OF THE INSTITUTE OF FUEL	1973-, Number 1-
THE ENGINEER	1952-, Volume 193- Weekly
ENGINEERING	1955-, Volume 179- Monthly
ENGINEERING AND PROCESS ECONOMICS Continued as ENGINEERING COSTS AND PRODUCTION ECONOMICS	1977-79, Volumes 2-4
ENGINEERING COMPUTERS	1984-, Volume 3- Bimonthly

Fig. 7 Journal catalogue

Almost certainly there will also have been other relevant books published which you don't know about and which are not in the library. If you have not found very much, you will probably need to extend your search by looking through lists of published books.

These lists, or bibliographies as they are called, vary considerably in the amount of information given about each title. The advantage of having some background knowledge is that you can adopt a fairly selective approach which will prevent you being overwhelmed with seemingly relevant material.

Current books
Someone working in a rapidly changing subject field or one in which there have been many recent developments – genetic engineering and information technology being obvious examples – will probably only be interested in books published within the last few years. If you are in this position you could begin by looking in *Whitaker's books in print*, which can be found in practically every library and information centre, however small.

When a book is said to be 'in print', that means that copies can be bought from its publisher. Since most books tend to stay in print for only a few years, the majority of titles included will be fairly recent. Older books listed are likely to be those which have proved to be of permanent value and have been reprinted when all the original copies were sold.

Whitaker's books in print arranges the titles, authors and editors in a single alphabetical sequence (Figure 8). The important information to note about any interesting book is the author(s) or editor(s), title, publisher, year of publication, price and the International Standard Book Number (ISBN). The price of a book usually gives some indication of its level. It would be reasonable, for instance, to assume that Hoelscher's *Graphics for engineers* at £57.20 is a fairly advanced work. The titles of books are normally inverted so that *Graphics for engineers*, for example, is also listed as *Engineers, Graphics for*.

It is worth mentioning at this point that many reference publications are now also produced in a computerized or electronic form as databases on a type of compact disc called CD-ROM. *Whitaker's books in print*, for example, is available in three different formats: as a printed reference book, on microfiche (as in Figure 8) and on CD-ROM, this version being known as *Bookbank CD-ROM*. Searching a database on CD-ROM is a fairly simple business – the CD-ROM disc is held in a player (similar to those used for playing audio compact discs) and searched using a microcomputer. The information found is displayed on a monitor (often in colour) and can be printed off if necessary. More will be said on the subject of CD-ROM in Chapter 4, but for now it is worth noting that searches made on a CD-ROM database are generally better and faster than those made in the printed equivalent. Not all that many libraries yet have CD-ROM versions; finance is a problem because of the cost of both the CD-ROM databases and the microcomputers needed to drive the systems. If, however, the library you are working in provides CD-ROM databases, try to use these where possible as it will make your searching much easier.

The books listed in *Whitaker's books in print* are the output of British publishers or of overseas publishers with bases in the UK. Inevitably, therefore, some foreign material will be missed. One good way to supplement your information is to check a list of American books to get the remainder of the output of the major English language publishers in science and technology. The equivalent US publication to *Whitaker's books in print* is *Subject guide to books in print*, up-to-date copies of which can usually be found in most reasonably sized libraries. The book titles are arranged under fairly broad subject categories, as the sample display in Figure 9 shows. Bowker, the publisher, has also produced a CD-ROM database called *Books in print PLUS* – well worth using if available.

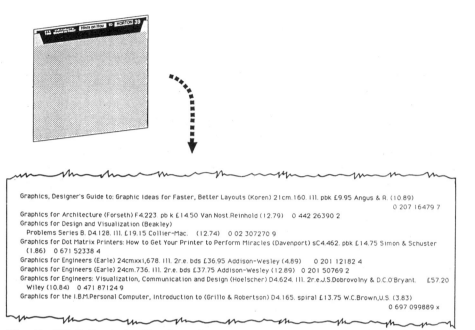

Graphics, Designer's Guide to: Graphic Ideas for Faster, Better Layouts (Koren) 21cm.160. Ill. pbk £9.95 Angus & R. (10.89)
    0 207 16479 7
Graphics for Architecture (Forseth) F4.223. pb k £14.50 Van Nost.Reinhold (12.79)   0 442 26390 2
Graphics for Design and Visualization (Beakley)
  Problems Series B. D4.128. Ill. £19.15 Collier-Mac.   (12.74)   0 02 307270 9
Graphics for Dot Matrix Printers: How to Get Your Printer to Perform Miracles (Davenport) sC4.462. pbk £14.75 Simon & Schuster
  (1.86)   0 671 52338 4
Graphics for Engineers (Earle) 24cmxxi,678. Ill. 2r.e. bds £36.95 Addison-Wesley (4.89)   0 201 12182 4
Graphics for Engineers (Earle) 24cm.736. Ill. 2r.e. bds £37.75 Addison-Wesley (12.89)   0 201 50769 2
Graphics for Engineers: Visualization, Communication and Design (Hoelscher) D4.624. Ill. 2r.e.J.S.Dobrovolny & D.C.O'Bryant.   £57.20
  Wiley (10.84)   0 471 871249
Graphics for the I.B.M.Personal Computer, Introduction to (Grillo & Robertson) D4.165. spiral £13.75 W.C.Brown,U.S. (3.83)
    0 697 099889 x

Fig. 8   Whitaker's books in print

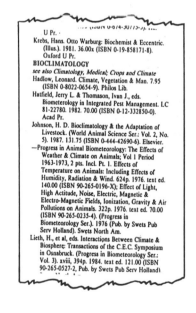

U Pr. ·
  ··· (ISBN 0-674-50775-5). ···
Krebs, Hans. Otto Warburg: Biochemist & Eccentric.
  (Illus.). 1981. 36.00x (ISBN 0-19-858171-8).
  Oxford U Pr.
BIOCLIMATOLOGY
see also Climatology, Medical; Crops and Climate
Hadlow, Leonard. Climate, Vegetation & Man. 7.95
  (ISBN 0-8022-0654-9). Philos Lib.
Hatfield, Jerry L. & Thomason, Ivan J., eds.
  Biometerology in Integrated Pest Management. LC
  81-22780. 1982. 70.00 (ISBN 0-12-332850-0).
  Acad Pr.
Johnson, H. D. Bioclimatology & the Adaptation of
  Livestock. (World Animal Science Ser.: Vol. 2, No.
  5). 1987. 131.75 (ISBN 0-444-42690-6). Elsevier.
—Progress in Animal Biometeorology: The Effects of
  Weather & Climate on Animals; Vol 1 Period
  1963-1973, 2 pts. Incl. Pt. 1. Effects of
  Temperature on Animals: Including Effects of
  Humidity, Radiation & Wind. 624p. 1976. text ed.
  140.00 (ISBN 90-265-0196-X); Effect of Light,
  High Actitude, Noise, Electric, Magnetic &
  Electro-Magnetic Fields, Ionization, Gravity & Air
  Pollutions on Animals. 322p. 1976. text ed. 70.00
  (ISBN 90-265-0235-4). (Progress in
  Biometeorology Ser.). 1976 (Pub. by Swets Pub
  Serv Holland). Swets North Am.
Lieth, H., et al, eds. Interactions Between Climate &
  Biosphere: Transactions of the C.E.C. Symposium
  in Osnabruck. (Progress in Biometeorology Ser.:
  Vol. 3). xviii, 394p. 1984. text ed. 121.00 (ISBN
  90-265-0527-2, Pub. by Swets Pub Serv Holland)

Fig. 9   Subject guide to books in print

Unless the subject is very new or very specialized, examining the British and American lists of current books should produce at least a handful of relevant titles. At this stage, you will need to decide whether to extend your search to cover older material or whether the books already discovered will be sufficient to be going on with. If you are looking for background information to familiarize yourself with a new topic, then a selection of recent material should be sufficient. If, on the other hand, the subject is central to a research project, then you may need to make a more extensive search of older books to be fairly certain that nothing has been missed.

Older books

Again it is probably best to concentrate first on tracing British books, following this up with a search for foreign titles, particularly those published in the USA. Since 1950, practically all books published in the UK have been recorded in an index called the *British national bibliography*, usually shortened to *BNB*. Books are not listed in alphabetical order by author, but by the Dewey Decimal Classification system, as the illustration in Figure 10 shows. Author, title and subject indexes are published in a separate volume.

For many years *BNB* only listed books which had already been published – its producers having access to the copies of books which publishers are legally bound to deposit with the British Library. Although the *BNB* is published in weekly issues, which are later cumulated into four-monthly and then annual parts, delays in processing meant that books could be anything between three and twelve months old before appearing in print. Nowadays many publishers supply information to *BNB* about their forthcoming books so that these can be listed before their actual publication date – an advantage if you want the newest books in the field. You can identify these books because they are marked 'CIP entry' on the record.

A good way of tracing American books is to look in a publication called the *Cumulative book index: a worldwide list of books in the English language* (New York, H. W. Wilson), a monthly publication with annual cumulations. Authors, titles and subjects are arranged in a single alphabetical sequence (Figure 11). Anyone needing to search back many years would also find the multi-volume *Pure and applied science books 1876 – 1982*, published by Bowker, very useful. There are several other lists of published books, but the ones we have outlined here should prove adequate for most searches.

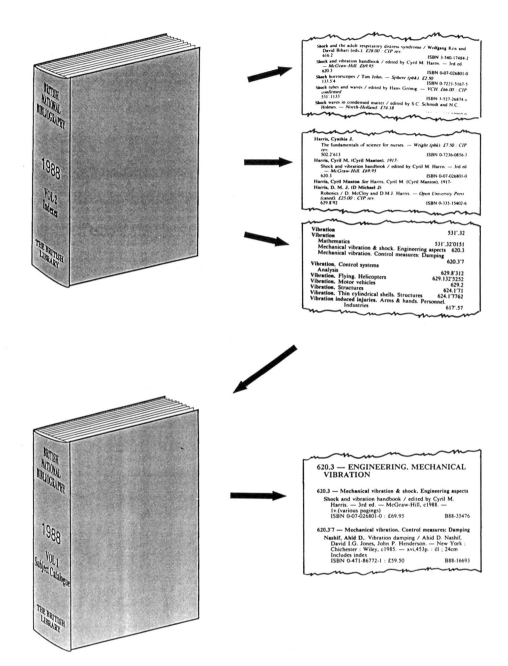

Fig. 10 *British national bibliography*

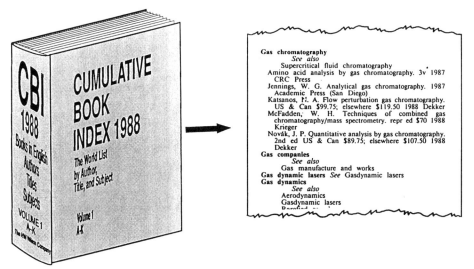

Fig. 11 *Cumulative book index*

Reviews
A good review found early on in a search can save a lot of time and effort. The problem with review articles is that they are scattered throughout many different types of literature – conference proceedings, books, theses, journals and review series – with the result that they can be difficult to find. A good place to start looking for reviews is the *Index to scientific reviews*, published twice a year by the Institute for Scientific Information (Figure 12). It does, despite its title, it also cover technical and medical disciplines. Review references are arranged in alphabetical order by name of author, and can be traced via the subject index – the link between keywords in the subject index and a reference being an author's name rather than a number.

Theses
Mention has already been made of the literature survey and review included in every thesis. Abstracts and indexes do not generally include many theses because of shortage of space, and *Index to scientific reviews* concentrates on journals and review publications. To find theses you need to use more specialized indexes. The best source of information on British theses is *Index to theses*, published four times a year by Aslib. These are arranged under broad subject headings, supported by subject and author indexes (Figure 13).

25

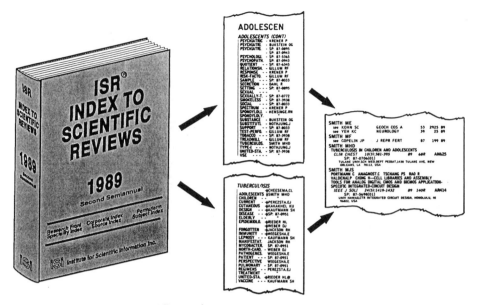

Fig. 12 *Index to scientific reviews*

Quite heavy use is made of the index by people who have heard of work carried out at a particular institution. Although they usually know roughly when the research was carried out, they sometimes have difficulties in finding the actual thesis. This is because there have been delays in the past between the submission of a thesis and its appearance in the index. Because of this it is a good idea when looking for an older thesis to check back through several years round about the expected date of submission.

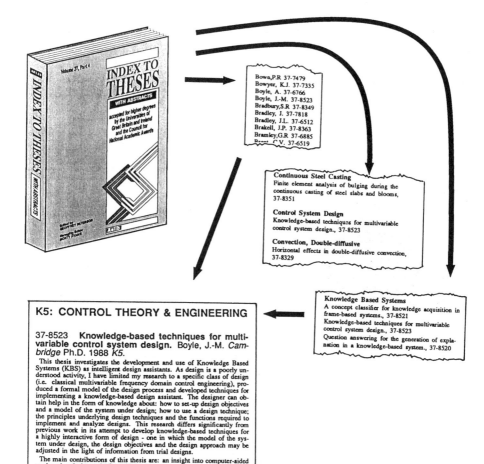

Fig. 13 *Index to theses*

Most US and Canadian theses can be found in *Dissertation abstracts international* (University Microfilms), Part B of which covers science and engineering (Figure 14). Searching back over a long period of time is possible with *Dissertation abstracts international* because University Microfilms have issued a series of cumulated indexes covering the years 1861 to 1987.

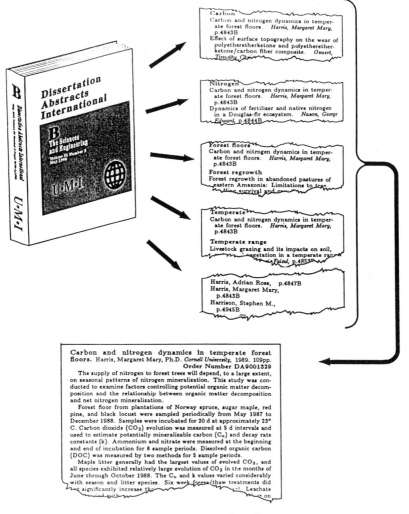

Fig. 14 *Dissertation abstracts international*

For a longish search back, however, it would be better to use the computerized version if possible.

Summary

- Find out as much as possible about the subject by following up the references in any papers or books in your possession.

- If you do not know of any previous work, follow up any references given about the subject in specialist encyclopedias or handbooks.

- Try one of the guides to published information to see if there is a section or chapter on the literature of the subject.

- Look for relevant books in your library or information centre, using the subject index and subject catalogue to make sure nothing is missed.

- Use lists of published books to look for other relevant publications.

- See if any reviews in journals, books, theses and so on, have been published.

3

Using abstracts and indexes

The importance of abstracts, indexes and their electronic counterparts, databases, in making the information contained in journals, conference proceedings and reports more accessible has already been mentioned in Chapter 1. For many years abstracts and indexes have been a vital requirement of any effective literature search, but with their widespread computerization as databases, the role of the printed publication has declined. Some knowledge of their structure and arrangement is still useful, though, both because there are circumstances where a manual search has to be made, and also because this is an aid in understanding the operation and use of the computerized equivalents.

Structure

The basic units of all abstracts and indexes, whatever the subject field, are descriptions of publications. In an index the descriptions consist only of references to publications – see for example *Index medicus* in Figure 15. In abstracts, as the name implies, the descriptions of publications also include summaries of the information contained, which is obviously an advantage in deciding if a publication is relevant or not. In some, such as *Chemical abstracts*, the summaries are quite lengthy, giving information about experimental methods as well as conclusions. In others the summaries are much briefer, only indicating the main findings. The layout used for abstracts and indexes does vary too, particularly in the way references are arranged. The basic elements, though, such as author, title and reference, can usually be very quickly identified, as the examples in Figure 15 show. It is perhaps worth noting here that *Engineering index,* despite its title, is actually an abstract.

Index medicus

Is poor pregnancy outcome a risk factor in rheumatoid arthritis? Spector TD, et al. **Ann Rheum Dis** 1990 Jan; 49(1):12–4

Chemical abstracts

113: 1425f **Structure of the human gene for α–enolase.** Giallongo, Agata; Oliva, Daniele; Cali, Larissa; Barba, Giovanna; Barbieri, Giovanna; Feo, Salvatore (Ist. Biol. Sviluppo, C. N. R., 90123 Palermo, Italy). *Eur. J. Biochem.* **1990,** 190(3), 567–73 (Eng). In mammals there are at $\geq$3 isoforms of the glycolytic enzyme enolase encoded by 3 similar genes: α, β, and γ. The human α–enolase gene was studied. The gene exists as a single copy in the haploid genome and is composed of 12 exons distributed over >18,000 bases. The structure of this gene has a high degree of similarity to that of the human and rat γ-enolase genes, with identical positions for all the intron regions. Primer extension and S1 nuclease protection expts. indicate that transcription is initiated at multiple sites. The putative promoter region, like that of other house–keeping genes, lacks canonical TATA and CAAT boxes, is extremely G + C–rich, and contains several potentia' SP1–binding sites. Furthermore, various sequences similar to known regulatory elements were detected.

Engineering index

071324 **Carbonyl iron powders. Its production and new developments.** Carbonyl iron powder (CIP) is a special iron powder with some outstanding properties. It is used in powder metallurgy (PM) as well as in the food, chemical and in the electronic industries. In recent years a new PM technique called metal injection moulding has been receiving more attention by designers, particularly for small and complex shaped PM parts. Because of its fineness and sphericity, CIP helped this new technology to become a promising technique. (Author abstract) 10 Refs.

Bloemacher, D.I. (BASF, Ludwigshafen, West Ger). *Met Powder Rep* v 45 n 2 Feb 1990 p 117-119.

Fig. 15 Sample abstracts

Arrangement

Abstracts and indexes which cover extensive subject areas or disciplines have to include large numbers of papers and consequently tend to be issued at frequent time intervals, often fortnightly or monthly. The giant of them all, *Chemical abstracts*, which includes nearly half a million papers a year, is published weekly. Frequent publication is in any event an advantage because it helps to cut down the delay between the publication of a paper in a journal or conference proceedings and its subsequent appearance in an abstract or index. Since any one issue of an abstract or index will contain hundreds of references to publications, some kind of subject arrangement is needed to help people find the information they want.

One way is to use a classification scheme, such as the one devised by the International Council for Scientific Unions for *Physics abstracts* (Figure 16). If a classification system is used, you can expect to find, as with *Physics abstracts*, a subject guide or index to the system in each issue to help you decide on the right number or code. Another common practice is to group references into fairly broad subject categories, as does, for example, *Biological abstracts* (Figure 17). A third way is to arrange references into very specific subject categories, with relatively few items in each category being listed out beneath the appropriate keywords or subject terms, a practice used by *Index medicus* (Figure 18). This type of arrangement avoids the need to scan through lots of irrelevant papers, but it is possible to miss useful information simply because it is listed under keywords you had not thought of.

Thesauri

One way in which the producers of this type of abstract and index try to get round this difficulty is by compiling a list, called a thesaurus, of the keywords or subject terms used for the arrangement of references. The thesaurus used by *Index medicus*, named *Medical subject headings* or *MeSH*, is shown in Figure 19. If a particular topic can be described by any one of several keywords or synonyms, a list such as this will indicate which one has been used in the arrangement, and link the keywords not chosen to the one which has been. The term 'PYKNOLEPSY', for example, is not used in *Index medicus*, hence the need for a reference elsewhere in the thesaurus 'PYKNOLEPSY see EPILEPSY, PETIT MAL'. A thesaurus will also indicate the links with different but related topics where there might be relevant information. Under the general heading 'EPILEPSY', for example, there is a reference to the related section 'CONVULSIONS'.

Subject guide

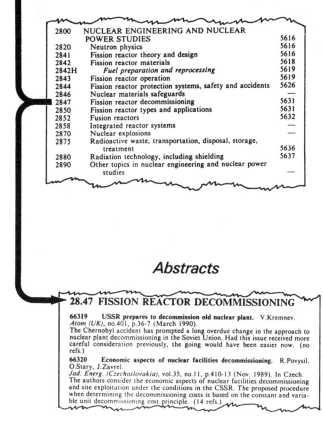

Fission breeder reactors	2850
Fission counters	2940
Fission-fusion reactor systems	2858
Fission power reactors	2850
Fission reactor decommissioning	2847
Fission reactor design	2841
Fission reactor fuel	2842
Fission reactor fuel preparation	2842H
Fission reactor fuel reprocessing	2842H
Fission reactor materials	2842
Fission reactor operation	2843
Fission reactor safety	2844
Fission reactor theory	2841
Fission reactor waste	2842
Fission research reactors	2850

Classification and contents

2800	NUCLEAR ENGINEERING AND NUCLEAR POWER STUDIES	5616
2820	Neutron physics	5616
2841	Fission reactor theory and design	5616
2842	Fission reactor materials	5618
2842H	*Fuel preparation and reprocessing*	5619
2843	Fission reactor operation	5619
2844	Fission reactor protection systems, safety and accidents	5626
2846	Nuclear materials safeguards	
2847	Fission reactor decommissioning	5631
2850	Fission reactor types and applications	5631
2852	Fusion reactors	5632
2858	Integrated reactor systems	—
2870	Nuclear explosions	—
2875	Radioactive waste, transportation, disposal, storage, treatment	5636
2880	Radiation technology, including shielding	5637
2890	Other topics in nuclear engineering and nuclear power studies	—

Abstracts

28.47 FISSION REACTOR DECOMMISSIONING

66319 **USSR prepares to decommission old nuclear plant.** V.Kremnev. *Atom (UK)*, no.401, p.36-7 (March 1990).
The Chernobyl accident has prompted a long overdue change in the approach to nuclear plant decommissioning in the Soviet Union. Had this issue received more careful consideration previously, the going would have been easier now. (no refs.)

66320 **Economic aspects of nuclear facilities decommissioning.** R.Povysil, O.Stary, J.Zavrel.
Jad. Energ. (Czechoslovakia), vol.35, no.11, p.410-13 (Nov. 1989). In Czech.
The authors consider the economic aspects of nuclear facilities decommissioning and site exploitation under the conditions in the CSSR. The proposed procedure when determining the decommissioning costs is based on the constant and variable unit decommissioning cost principle. (14 refs.)

Fig. 16 Physics abstracts

MAJOR CONCEPT HEADINGS FOR ABSTRACTS

Public Health AB-906
Radiation Biology AB-953
Reproductive System AB-957
Respiratory System AB-975
Sense Organs, Associated Structures and
    Functions AB-988
Social Biology (includes Human Ecology) AB-1010
Soil Microbiology AB-1011
Soil Science AB-1013
Temperature: Its Measurement, Effects and
    Regulation AB-1027
Tissue Culture: Apparatus, Methods and
    Media AB-1028
Toxicology AB-1029
Urinary System and External Secretions AB-1064
Veterinary Science AB-1074
Virology, General AB-1075

Biol Abstr 89(11):AB-1010

SOCIAL BIOLOGY
(includes Human Ecology)

See also: *Aerospace and Underwater Biological Effects – Ecology and Psychology • Behavioral Biology – Human Behavior • Gerontology • Psychiatry • Public Health*

121191. HELFENSTEIN, ULRICH. (Biostatistical Cent. Med. Dep., Univ. Zurich, Plattenstrasse 54, CH–8032 Zurich, Switzerland.) ACCID ANAL PREV 22(1): 79–88. 1990. Whe aid a reduced speed limit show an effect? Exploratory identification of an interventio. time.—In a statistical analysis of accident data before and after a speed limit reduction, the time of the countermeasure is, of course, well known. Our understanding of the accident process may, however, be increased if we assume in a thought experiment that this time is unknown. We ask if the data themselves can tell us something about such a possible time. By means of time series of traffic accidents in Zurich before and after a speed limit reduction, different exploratory methods are presented to identify the "unknown" time of this measure. For most of the investigated series, the most likely time was found to lie in the three months before the true introduction. A possible explanation of this result may be that the media already informed the public before the countermeasure was actually introduced. This finding leads to an improved parsimonious intervention model which distinguishes between a possible "pre-intervention effect" and the usual "intervention effect".

121192. GOLOB, THOMAS F., WILFRED W. RECKER and DOUGLAS W. LEVINE. (Inst. Transportation Studies, Univ. Calif. Irvine, Irvine, Calif. 92717, USA.) ACCID ANAL PREV 22(1): 19–34. 1990. Safety of freeway median high occupancy vehicle lanes: A comparison of aggregate and disaggregate analyse~
address~   ~ith High Occup~

Fig. 17 *Biological abstracts*

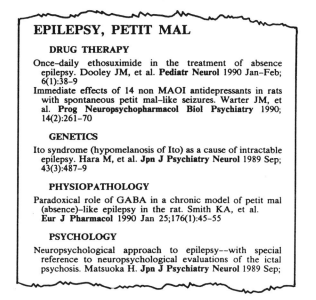

EPILEPSY, PETIT MAL

DRUG THERAPY

Once–daily ethosuximide in the treatment of absence epilepsy. Dooley JM, et al. **Pediatr Neurol** 1990 Jan–Feb; 6(1):38–9
Immediate effects of 14 non MAOI antidepressants in rats with spontaneous petit mal–like seizures. Warter JM, et al. **Prog Neuropsychopharmacol Biol Psychiatry** 1990; 14(2):261–70

GENETICS

Ito syndrome (hypomelanosis of Ito) as a cause of intractable epilepsy. Hara M, et al. **Jpn J Psychiatry Neurol** 1989 Sep; 43(3):487–9

PHYSIOPATHOLOGY

Paradoxical role of GABA in a chronic model of petit mal (absence)–like epilepsy in the rat. Smith KA, et al. **Eur J Pharmacol** 1990 Jan 25;176(1):45–55

PSYCHOLOGY

Neuropsychological approach to epilepsy––with special reference to neuropsychological evaluations of the ictal psychosis. Matsuoka H. **Jpn J Psychiatry Neurol** 1989 Sep;

Fig. 18 *Index medicus*

Subject indexes

Although a good subject arrangement is a useful aid in scanning through a few issues of an abstract or index, it is too slow and imprecise an instrument when searching for information published over a time-scale of several years. What is needed for this type of search is a detailed subject index to all the publications included. Those indexes and abstracts which have adopted a broad subject categorization or classification system often publish a subject index in each issue. Both *Chemical abstracts* and *Biological abstracts* do this, and it is very useful because it saves you the effort of scanning through a mass of what is mostly irrelevant or fringe interest material.

Generally speaking, most abstracts and indexes produce a cumulated subject index either once or twice a year to all the separately produced issues. These are obvious time-savers because you have only one index to check instead of perhaps six or twelve separate ones. Some of the larger abstracts and indexes also produce cumulated subject indexes covering several years; the five-yearly subject indexes to *Chemical abstracts* are famous for their size. INSPEC, the publisher of *Physics abstracts* and the complementary *Electrical & electronics abstracts* and *Computer & control abstracts*, produces four-yearly cumulations of its subject indexes. Such cumulated subject indexes can be expensive, though, and restricted funding and the trend towards computer-ized searching (which will be discussed in Chapter 4) have meant that libraries and information centres are now less likely to buy such cumulations

EPILEPSY
C10.228.140.490+

see related
 KINDLING (NEUROLOGY)
XR CONVULSIONS

EPILEPSY, ABDOMINAL see EPILEPSY, TEMPORAL LOBE

EPILEPSY, FOCAL
C10.228.140.490.207+

77; was indexed under JACKSONIAN SEIZURE 1970–76
X   EPILEPSY, PARTIAL

EPILEPSY, GRAND MAL
C10.228.140.490.234.225

EPILEPSY, MYOCLONUS
C10.228.140.490.234.248

77
X   LAFORA DISEASE
X   LUNDBORG-UNVERRICHT SYNDROME
X   MYOCLONIC EPILEPSY, PROGRESSIVE

EPILEPSY, PARTIAL see EPILEPSY, FOCAL

EPILEPSY, PARTIAL COMPLEX see EPILEPSY, TEMPORAL LOBE

EPILEPSY, PETIT MAL
C10.228.140.490.234.270
X   ABSENCE
X   AKINETIC PETIT MAL
X   PYKNOLEPSY

EPILEPSY, PSYCHIC EQUIVALENT see EPILEPSY, TEMPORAL LOBE

EPILEPSY, PSYCHOMOTOR see EPILEPSY, TEMPORAL LOBE

EPILEPSY, TEMPORAL LOBE
C10.228.140.490.207.301

64
X   EPILEPSY, ABDOMINAL
X   EPILEPSY, PARTIAL COMPLEX
X   EPILEPSY, PSYCHIC EQUIVALENT
X   EPILEPSY, PSYCHOMOTOR
X   EPILEPSY, UNCINATE

EPILEPSY, TRAUMATIC
C10.228.140.490.283   C21.866.460.152.411

EPILEPSY, UNCINATE see EPILEPSY, TEMPORAL LOBE

Fig. 19 *Medical subject headings (MeSH)*

than in the past. An alternative system is that used by abstracts and indexes such as *Index medicus*. This arranges the references and abstracts according to very specific subject categories in each issue; it does not produce separate subject indexes, but rather cumulates all the references annually into one complete sequence.

Methods of subject indexing

Superficially at least, all subject indexes look the same: a list of keywords and numbers linking up to the references. In fact there are two contrasting methods of subject indexing, producing two quite different types of index.

If you are to get good results when searching abstracts or indexes, you need to know what method of subject indexing has been used. In the first method, called controlled vocabulary indexing, publications are indexed using words chosen from a thesaurus (the list of suitable, approved keywords or subject terms we mentioned earlier). Many abstracts and indexes, including *Metals abstracts*, *Physics abstracts* and its two sister publications, *Computer & control abstracts* and *Electrical & electronics abstracts*, are indexed in this way. An example of *The index guide*, the list used by *Chemical abstracts*, is shown in Figure 20. Anyone using *Chemical abstracts* should check their choice of keywords in *The index guide* before starting. Someone looking for information on 'Paracetamol', for example, would find no papers listed in the subject indexes because this is listed under 'Acetamide, N– (4-hydroxyphenyl)–'. Note that *Chemical abstracts* give each substance a unique identifier, known as a Registry Number, the one for paracetamol being [103-90-2].

In the other method, called natural language indexing, papers are indexed using words from the title and text. This may not seem significantly different from indexing using a thesaurus, but it means that synonyms are separated in the index. For example, a paper with a title such as 'The effect of alcoholism on the human body' would be indexed under 'alcoholism';

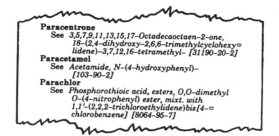

Paracentrone
  See *3,5,7,9,11,13,15,17–Octadecaoctaen-2-one,*
        18–(2,4–dihydroxy-2,6,6–trimethylcyclohexy=
        lidene)–3,7,12,16–tetramethyl– [31190-20-2]
Paracetamol
  See *Acetamide, N–(4-hydroxyphenyl)–*
        [103-90-2]
Parachlor
  See *Phosphorothioic acid, esters, O,O–dimethyl*
        O–(4–nitrophenyl) ester, mixt. with
        1,1'–(2,2,2–trichloroethylidene)bis[4–=
        chlorobenzene] [8064-95-7]

Fig. 20 *Chemical abstracts index guide*

a paper entitled 'The effect of excess drinking on the body', which did not mention alcoholism, would be indexed under 'excess drinking'. A subject term such as 'database', which is sometimes hyphenated, would appear in both forms: 'database' and 'data-base'. The singular and plural versions of a term, for example, 'crop' and 'crops', would also be separated. What this means in practice is that if you are using this type of index you will need to look at all variant spellings and synonyms to be certain of not missing anything.

Other types of index

Subject indexes, although very important when searching for information, do need supplementing by other types of index. Author indexes in particular are necessary to allow searches for the publications of individual scientists and technologists. Some, such as *Metals abstracts and index* (Figure 21) give the abstract number after the author's name. Others, such as *Physics abstracts*, also give the title of the paper.

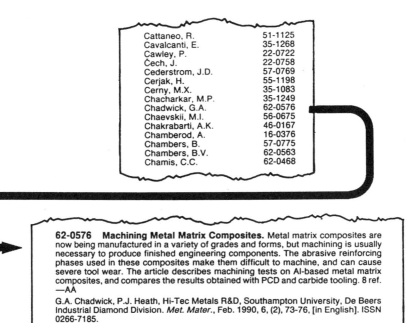

Fig. 21 *Metals abstracts and index*

39

Anyone trying to trace a patent may find the patent index in *Chemical abstracts* useful. This lists all patents included in *Chemical abstracts* in numerical order by country and can be used to trace an English language version of a foreign patent. Other abstracts and indexes also contain specialized indexes. *Biological abstracts*, for instance, has both a Generic and a Bio-systematic index in addition to its general subject index.

There are no really hard and fast rules which can be laid down as to what kind of indexes you are likely to find in the publication you use. The important thing is to be aware of the existence of these specialized types of index and to check them to see if they would be useful for your type of search.

Choosing an abstract or index

An important aspect of any literature search is the choice of a suitable abstract or index. How can someone new to literature searching select the best one for their topic? The abstracts and indexes mentioned so far in this chapter – *Chemical abstracts*, *Engineering index*, *Index medicus*, *Biological abstracts*, *Metals abstracts* and those published by INSPEC: *Computer & control abstracts*, *Physics abstracts* and *Electrical & electronics abstracts* – are large general services. They cover, between them, most scientific and technical subjects but there are also more specialized services which could prove equally useful.

The large number of abstracts and indexes published make it impossible for us to give a complete inventory. What we have done is to draw up a list of a few of the most widely available and most-used ones (Table 1). All of these publications are also available in electronic form as databases. In this format they are frequently given a different name to distinguish them from the printed version – Table 1 shows the database names in column 2. There are, of course, publications which contain exhaustive lists of all the abstracts and indexes currently in existence. One which can be found in most libraries is *Ulrich's international periodicals directory*, published annually by Bowker. Another approach is to ask the staff in your library or information centre what abstracts and indexes they actually have immediately to hand which cover the particular subject field. The stock of a library is obviously chosen to reflect the subject interests of the organization it serves. If a subject field has been of interest to the organization for some time it is likely that the library or information centre will have bought the specialized abstracts and indexes for the subject.

It may well be that you will be faced with a choice between two or three abstracts and indexes, particularly if you are working in a subject field on the fringe of several disciplines. Toxicology, for instance, is generally considered to be a branch of medical science, but the study of the toxic effects of chemicals on plants and animals is also of interest to biologists and chemists. Because of this, toxicological literature is included in both *Chemical abstracts* and *Biological abstracts*, in addition to *Index medicus*.

Table 1 Abstracts, indexes and databases

Abstract/index name	Database name(s)	Producer(s)
Biological abstracts	BIOSIS PREVIEWS	Biosciences Information Service
CAB (range of abstracts in agricultural sciences)	CAB Abstracts	CAB International
Chemical abstracts	CA Search/ CAS Online	Chemical Abstracts Service
Computer & control abstracts	INSPEC	Institution of Electrical Engineers (IEE)
Electrical & electronics abstracts	INSPEC	IEE
Engineering index	COMPENDEX*PLUS	Engineering Information
Excerpta medica	EMBASE	Elsevier Science
Government reports announcements and index	NTIS	National Technical Information Service
Index medicus	MEDLINE	National Library of Medicine
INIS atomindex	INIS	International Atomic Energy Agency
ISMEC: mechanical engineering abstracts	ISMEC	Cambridge Scientific Abstracts
Mathematical reviews	Math Sci	American Mathematical Society
Metals abstracts	METADEX	Institute of Metals/ American Society of Metals
Physics abstracts	INSPEC	IEE
Science citation index	SCI SEARCH	Institute for Scientific Information
World patents index gazette service	WPI	Derwent Publications

If you find yourself in this situation, there are two factors which are worth bearing in mind when deciding which abstract or index to use. The first and most important is the type of subject information which has been included. Although a subject field may be covered by several abstracts and indexes, the emphasis in each will tend to be different. To go back to our example of toxicology, it would be reasonable to expect that *Index medicus* would concentrate on covering toxicological information from the medical angle, and that *Biological abstracts* would emphasize the effect of toxins on the environment. Just browsing through a couple of issues of each publication should be enough to give you a fairly good idea of the type of information included. It is also worth looking to see if the publishers of an abstract or index issue a policy statement about their subject coverage. Many organizations do state their policy somewhere amongst the preliminary information, although you may have to hunt around to find it. Most abstracts and indexes also publish a list, such as the one produced by INSPEC shown in Figure 22, of the journals which are scanned for suitable papers. If you know of or regularly browse through several journals in the subject field, it is also worth checking to see if these are covered by the publication.

Comput. Humanit. (Netherlands)
Computers and the Humanities Kluwer Academic Publishers Group, P.O.Box 322, 3300 AH Dordrecht, Netherlands. From Jan.-March 1984 to 1987 published in the USA [Prior to 1988: Comput. & Hum. (Netherlands)]

Comput. Humanit. (USA)
Computers and the Humanities Paradigm Press Inc., P.O.Box 1057, Osprey, FL 33559-1057, USA. Prior to Jan.-March 1984 and since 1988 published in the Netherlands [Prior to 1988: Comput. & Hum. (USA)]

Comput. Ind. (Netherlands)
Computers in Industry North-Holland Publishing Co., P.O. Box 103, 1000 AC Amsterdam, Netherlands.

Comput. Ind. Eng. (UK)
Computers & Industrial Engineering Pergamon Press Ltd., Headington Hill Hall, Oxford OX3 0BW, UK. [Prior to 1988: Comput. & Ind. Eng. (GB)]

Comput.-Integr. Manuf. Syst. (UK)
Computer-Integrated Manufacturing Systems Butterworth Scientific Ltd., P.O.Box 63, Westbury House, Bury Street, Guildford, Surrey GU2 5BH, UK.

Comput. Intell. (Canada)
Computational Intelligence National Research Council of Canada, Ottawa, Ont. K1A 0R6, Canada.

Comput. J. (UK)
Computer Journal Published for the British Computer Society by Cambridge University Press, The Pitt Building, Trumpington Street, Cambridge CB2 1RP, UK. [Prior to 1988: Comput. J. (GB)]

Comput. Lang. (UK)
Computer Languages Pergamon Press Ltd., Headington Hill Hall, Oxford, OX3 0BW, UK. [Prior to 1988: Comput. Lang. (GB)]

Comput. Lang. (USA)
Computer Language Miller Freeman Publications, 500 Howard Street, San Francisco, CA 94105, USA.

Comput. Law (UK)
Computers and Law Kluwer Law, 1 Harlequin Avenue, Brentford, TW8 9EW, UK.

Comput. Methods Appl. Mech. Eng. (Netherlands)
Computer Methods in Applied Mechanics and Engineering North-Holland Publishing Co., P.O. Box 211, 1000 AE Amsterdam, Netherlands. [Prior to 1988: Comput. Methods Appl. Mech. & Eng. (Netherlands)]

Comput. Methods Programs Biomed. (Netherlands)
Computer Methods and Programs in Biomedicine Elsevier Science Publishers B.V. (Biomedical Division), P.O.Box 1527, 1000 BM Amsterdam, Netherlands. [Prior to 1988: Comput. Methods & Programs Biomed. (Netherlands)]

Comput. Music J. (USA)
Computer Music Journal MIT Press, 55 Hayward Street, Cambridge, MA 02142, USA.

Comput. Negot. Rep. (USA)
Computer Negotiations Report International Computer Negotiations Inc., 238 Christopher Street, Upper Montclair, NJ 07043, USA. [Prior to 1988: Comput. Negotiations Rep. (USA)]

Comput. Netw. ISDN Syst. (Netherlands)
Computer Networks and ISDN Systems Elsevier Science Publishers B.V., P.O.Box 211, 1000 AE Amsterdam, Netherlands. [Prior to 1988: Comput. Networks & ISDN Syst. (Netherlands)]

Comput. News (UK)
Computer News CW Communications, 99 Gray's Inn Road, London WC1X 8UT, UK. [Prior to 1988: Comput. News (GB)]

Comput. Newsl. Sch. Bus. (USA)
Computing Newsletter for Schools of Business College of Business Administration, University of Colorado, Austin Bluffs Parkway, Colorado Springs, CO 80907, USA. [Prior to 1988: Comput. Newsl. Schools Bus. (USA)]

Comput. Newsp. (UK)
Computing, The Newspaper VNU Business Publications BV, 55 Frith Street, London W1A 2HG, UK. Supersedes in part: Computing [Prior to 1988: Comput. Newspap. (GB)]

Fig. 22 INSPEC list of journals

42

The second point to consider is what forms of literature have been included. Most of the larger abstracts and indexes include conference proceedings as well as journals, but some of the more specialized ones restrict themselves to journals. Even large abstracts and indexes can be very selective when it comes to including reports, books and theses, and with the exception of *Chemical abstracts* most exclude patents completely. If you are specifically looking for the type of information found in patents, you are probably best advised to use the specialized patent indexes and abstracts such as the *World patents index gazette service*. If you suspect that quite a lot of work in the subject field is written up in report form rather than as papers in journals, then it may also be worth searching a publication which specifically abstracts these, such as *Government reports announcements and index*.

Hints on searching
Although the following points may seem obvious, they are often overlooked by people when using abstracts and indexes. Most people can spend only a limited amount of time on literature searching so it is important to use it as effectively as possible.

- Some information is generally given in an abstract or index on its use, and frequently a sample reference is included with the separate parts, for example author, title and so on, identified. Do read or at least glance through such advice before starting, if only to ensure that there are no peculiarities which you have not come across before.

- Check all the separate issues and bound volumes of the abstract or index, so that you know what indexes have been published and which years these cover. If the subject and author indexes for a particular year and the separate issues to which these refer are not too bulky, these may all have been bound in one volume. If the subject and author indexes are large, these may all have been bound separately from the associated issues. Where two sets of subject and author indexes are produced each year, that is, at six-monthly intervals, these might be bound together into one joint volume even though they refer to separately numbered volumes of abstracts. Practices vary between libraries – the important thing is to check to make certain you have not missed anything.

- Draw up a list of all the keywords which you think might possibly have been used to describe the topic and check all these in the most recent index, or better still in the thesaurus (list of keywords used), if there is one. Follow up any/see or *see also* references to related subject fields to see if any of the items are relevant. Once you have done this you will have a good idea which keywords will be the most productive to use. As new topics are introduced and subject fields develop, abstracts and

43

indexes obviously introduce new keywords into their indexes; if you intend to search back over several years you need to keep a careful check on the keywords used to make sure that these have not changed.

- Begin your search with the most recent issue of the abstract or index and work backwards from this date. There are two reasons why it is better to do this rather than working forwards in time from a certain year. First you could find a review in a fairly recent issue. If this is central to your topic then most of the previous work will have been covered and you will not need to search any further back. Secondly, if there is not enough time to go back as far as you would have liked you will at least have the most current papers, and the references in these will take you further back in time.

- Keep a record of *all* essential information on any relevant references you find. This may sound very obvious but it is tempting, especially if you are in a hurry, to leave out the title of the paper or perhaps the volume or issue of the journal. If you find a reasonably large number of references, say 30 or 40, you may only want to look at the 10 or so most relevant to start with, but without the titles of the papers you may not be able to rank them in order of priority. Volume numbers are important as a safeguard against inaccurate references. If, for instance, the year of publication has been copied down wrongly, then it should still be possible with a volume number to find the correct paper. Try also to resist the temptation to abbreviate drastically the titles of journals. There are, for example, at least four journals with titles beginning 'High temperature' – *High temperature, High temperature science, High temperature technology* and *High temperatures – high pressures* – If you abbreviate to *High temp*, who will know what you mean?

- A good precaution against mistakes is to keep a note of the year and abstract number of each reference; CA 113:1425f, for example, would mean the reference was found in *Chemical abstracts*, volume 113, at abstract number 1425f.

- If, after searching back through several issues, it looks as though you will get either too many or too few references, modify your search technique or strategy. If there are too many references, you may have to be more selective in your choice of keywords. It is unlikely that you will have sufficient time to follow up every item, find the original document and read it, so it is important to evaluate each reference carefully at the selection stage. While there is always a chance of missing some useful information by ignoring a reference, experience shows that relevant publications will probably be cited by other workers. You should become aware of them by reading the papers you do follow up from

44

your initial search. If you think you are finding too few references, then you may need to broaden your choice of keywords and be less selective in your evaluation of each reference.

Science citation index

Some of the ways of building upon information already known have been outlined in the previous chapter. By this stage most people are likely to have found at least one paper of importance to their project. Other researchers will almost certainly have drawn upon the work in this paper and will consequently have included (or cited) the paper in their list of references. It is possible to find out who has subsequently referred to this older paper by using a product called *Science citation index* or *SCI*.

SCI differs radically from the abstracts and indexes described so far, and to use it effectively it is necessary to appreciate how it is compiled and organized. The basic principle behind the *Index* rests on the fact that although more than 30,000 scientific and technical journals are published worldwide each year, most of the significant information is contained within a much smaller number. The policy of the Institute for Scientific Information, the producers of *SCI*, is to index the contents of selected journals (approximately 3,200 in 1990) in science and technology. Every paper in these journals is included (whatever the subject field) along with all the references or citations to publications made by the authors to previous work. The direct linkage of these earlier references to recently published papers then makes it feasible to carry out a literature search 'forward in time' using the earlier reference as the starting-point. *SCI* is marketed both as a printed publication and on CD-ROM; the example which follows, illustrating how recent references to a specific paper can be traced, utilizes the CD-ROM version.

Suppose, for instance, you were interested in information on superconductivity at high temperatures and had, as a starting-point, the paper by Bednorz *et al.* published in *Europhysics letters* and shown in Figure 23. As can be seen in Figure 24a, when the name of a known author is entered into the system it will, given the appropriate commands, display a list of all papers by that author which have been cited or referred to in any of the 3,200 journals indexed by *SCI* between January and June 1990. After selection of the paper in *Europhysics letters*, the system then displays the number of citations made (in this case seven) to the paper in the first six months of 1990 (Figure 24b). The first of these more recent citations is then shown in Figure 24c. More references to Bednorz's original paper could be found by repeating the search using the earlier years of the *SCI* database, that is, 1989, 1988 and 1987.

EUROPHYSICS LETTERS

Europhys. Lett., 3 (3), pp. 379-385 (1987)

1 February 1987

Susceptibility Measurements Support High-T_c Superconductivity in the Ba-La-Cu-O System.

J. G. Bednorz, M. Takashige(*) and K. A. Müller

IBM Research Division, Zurich Research Laboratory
CH-8803 Rüschlikon, Switzerland

(received 22 October 1986; accepted 12 November 1986)

PACS. 74.70. – Superconducting materials.
PACS. 74.10. – Occurrence, critical temperature.
PACS. 74.70N – Dirty superconductors.

Abstract. – The magnetic susceptibility of ceramic samples in the metallic BaLaCuO system has been measured as a function of temperature. This system had earlier shown characteristic sharp drops in resistivity at low temperatures. It is found that, for small magnetic fields of less than 0.1 T, the samples become diamagnetic at somewhat lower temperatures than the resistivity drop. The highest-temperature diamagnetic shift occurs at (33 ± 2) K, and may be related to shielding currents at the onset of percolative superconductivity. The diamagnetic susceptibility can be suppressed with external fields of 1 to 5 T.

Fig. 23   Starting paper for a citation search

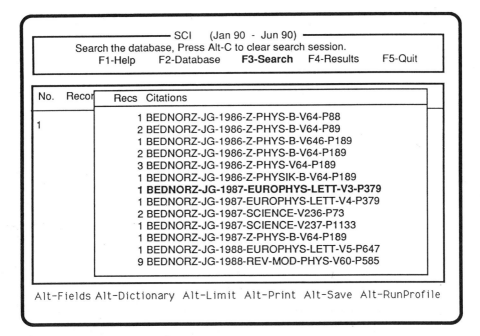

Fig. 24a   Screen display from *Science citation index*, compact disc edition

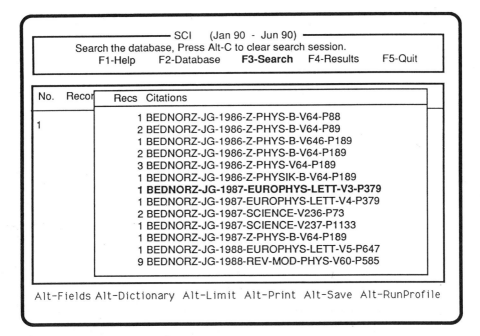

Fig. 24b   Screen display from *Science citation index*, compact disc edition

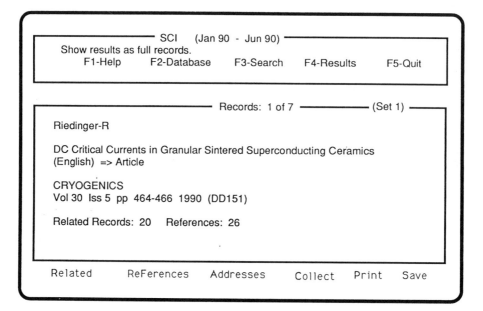

Fig. 24c Screen display from *Science citation index*, compact disc edition

Summary

- When choosing an **abstract** or index, look at the way the subject is emphasized. See what types of literature are included: journals, conferences, reports and so on. Check the list of journals indexed for ones relevant to your subject.

- Examine the subject indexes to see if these have been compiled using a thesaurus, or if the keywords have been taken straight from the title and text.

- Work backwards in time from the most recent issue and modify your search technique if you are getting either too much or too little information.

- Keep an accurate record of references.

- Check any really important reference in *Science citation index* to see if it has since been cited by other authors.

4

Computerized information searching

The benefits of using computers to search electronic versions of reference books, encyclopedias and so on have already been briefly mentioned in Chapter 2. However, it is the use of computers to search the electronic versions of abstracts and indexes that has resulted in the greatest benefits so far in literature searching. Virtually all abstracts and indexes are now produced on magnetic tape as well as in the form of printed publications.

However, by itself a database on magnetic tape is not particularly useful since all the information is in linear format. This means that all of the records on a tape have to be searched in sequence in order to obtain information about perhaps just one or a few items. In the early years of computerized information retrieval, literature searches were carried out in just this way. This method of searching has been replaced by 'online searching', in which the information content of abstracts and indexes is stored not on magnetic tape but on either magnetic or CD-ROM discs. Magnetic discs would normally be held on a mainframe computer and accessed via a telecommunications link. CD-ROM discs, which are becoming increasingly common, can be read by a CD-ROM player or drive and searched using a microcomputer, thus requiring no telecommunications link. The obvious advantage of discs, either magnetic or CD-ROM, as storage media is that a single piece of information can be accessed without the necessity of searching through all the other items in the file. The discs containing the databases can then be directly searched by the computer and questions about the information contained in the files can be answered immediately. If the enquirer is dissatisfied with the result – maybe the information is not relevant, or perhaps there is too little or too much – then the question can be reformulated immediately and the process repeated until a better answer is obtained. This in essence is how computerized information searching operates although in practice, of course, more procedures are involved.

Online systems

The producers of abstracts and indexes, although the sources of the databases used in the systems, do not generally provide computerized searching systems themselves. Instead they lease the magnetic tapes to other organizations operating powerful mainframe computers with large storage capacities.

These organizations, usually referred to as hosts, load the data held on magnetic tape on to the magnetic discs in such a way that information about names of authors, titles of papers and the keywords or subject terms describing the content can be retrieved instantly. Developing the advanced software programs needed for computerized searching is expensive, but once in operation the same software can be used for searching many different databases. For this reason most hosts provide access to as many databases as possible. For example, DIALOG Information Services, Inc., one of the major hosts, had around 380 databases available for searching in 1990. DIALOG^R itself is based in California, but there are other important hosts located in Europe. Notably these include DATASTAR, based in Switzerland, and ESA-IRS, based at the European Space Agency in Italy.

Whatever the location of the host, however, all these types of system operate in a basically similar fashion: a terminal, usually a microcomputer, is connected over a long-distance telecommunications link to the host's computer, when it is said to be online (Figure 25). The terminal is usually connected to the telecommunications network by a modem or acoustic coupler. It is possible to communicate directly with a host's computer simply by dialling the appropriate telephone number, but unless this is located a fairly short distance away, the telecommunications costs incurred are prohibitive. For this reason special data communication networks are normally used; such networks, which operate across continents, provide fast, reliable transmission, independent of distance and at relatively low cost. A terminal is connected to such a network by dialling its nearest access point, called a node. Once connected to the host's computer, information about the search can be input using the keyboard, the responses from the computer being displayed on the monitor.

The systems are operated on a time-sharing basis, with thousands of terminals being connected to a host's computer at any moment. A response time of a few seconds is fairly average but can stretch to well over a minute at peak times of the day. The information found during a search – references to papers in journals, conference proceedings and so on – can be printed out online at the terminal, provided, of course, that a printer is attached. However, online printing is relatively slow and expensive because of the telecommunications costs, and tends to be used only where the references are needed urgently or when only a small number of references are produced.

52

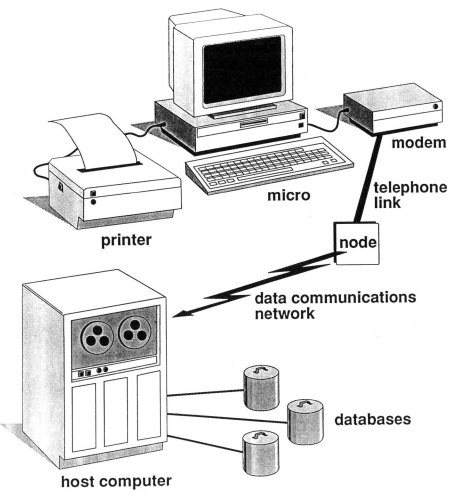

Fig. 25   An online searching system

Alternatively, references can be printed out by the host's computer offline, generally arriving a few days after a search. When a microcomputer is used as a terminal it is also possible to transmit the references over the telecommunications link, a process known as downloading. Permission is required from the database producer and host; provided this is obtained, the references can then be printed out on site.

Payment for the use of the system depends on a combination of factors including:

- Which database is used. Some databases are much more expensive than others.

- How long the search takes. The longer the time spent connected online, the higher the costs will be.

- The number of references produced.

As a very rough guide, the cost of an online search works out at around £1 for every minute the searcher is connected online. For a search taking 10 to 15 minutes (a fairly typical time) the price could work out at about £40 to £50 when the cost of printing out references is added on.

Because of the relatively high cost, these types of systems have, at least until recently, been mainly used by information professionals carrying out searches on behalf of their readers. This situation is now changing and more so-called 'end-users', that is, the people who actually need the information, are starting to use the systems.

CD-ROM systems

During the latter half of the 1980s an alternative method of computerized searching using CD-ROM optical discs has been developed. CD-ROM discs, as mentioned in Chapter 2, can be used to store information in electronic or digital form. Because CD-ROM has a high storage capacity it is possible to 'publish' a disc containing all or part of a database. The amount of information which can be held on one CD-ROM obviously depends on the size of a database. For example, one commercially available CD-ROM version of MEDLINE (the computerized version of *Index medicus*) contains the equivalent of six months of the printed copy (six volumes containing in total 7,938 pages). A smaller database would obviously be able to store proportionately more of the printed version on one CD-ROM. The software used to search CD-ROM discs is similar in many respects to that used in online searching, so that the same kind of search strategies can be adopted and comparable performances achieved. This software, which is supplied with the CD-ROM disc, is usually installed on to the hard disc of the microcomputer so that the discs are immediately searchable. Quite a large number of databases are now also 'published' on CD-ROM – MEDLINE, COMPENDEX*PLUS (the computerized version of *Engineering index*) and *Biological abstracts on compact disc*, for example. A typical CD-ROM system is shown in Figure 26.

As mentioned earlier, one advantage of using CD-ROM discs for computerized searching is that the systems are 'stand-alone', requiring no telecommunications link to a distant host computer. In addition, CD-ROM

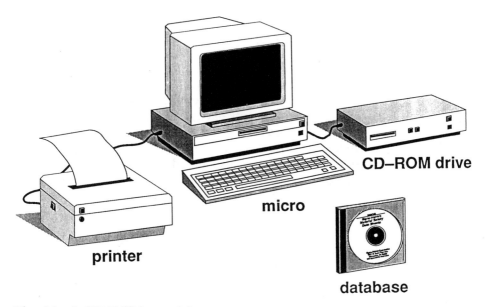

Fig. 26 A CD-ROM searching system

databases are either leased or bought outright by a library for a set amount – once purchased they can be used as much as needed without any further cost. Because of this, CD-ROM searching is becoming very popular with end-users. The systems are easy to use – some help may initially be needed from library staff but it should be possible after that for people to carry out their own search quickly and simply.

Search methods – manual, online or CD-ROM?

With three separate methods of searching available – manual searching of printed abstracts and indexes, online searching via hosts and local searches of CD-ROM databases – how can a newcomer decide which method would be the best to follow? To answer this question we need to look at the main benefits and problems of each method.

Perhaps the most obvious benefit of computerized searching, whether online or CD-ROM, is speed. A search back through five or ten years of a printed abstract or index might take at least half a day, whereas a search through the equivalent computerized version could take half an hour or less. In addition, the references obtained can be printed out in full after the search. All the tedium of copying down the information on to paper, with all the inaccuracies which can creep in, is eliminated completely.

Another major advantage of computerized searching over manual methods

55

is that it is possible to carry out a more effective search. To understand why this is so, it is necessary to look at how references are indexed and stored in databases. Compare, for example, the way the reference in Figure 27 appears in *Electrical & electronics abstracts* and in the INSPEC database (of which *Electrical & electronics abstracts* forms a part). The reference in the printed version has been indexed using two compound keywords: *flat panel displays* and *liquid crystal displays*, both selected from the *INSPEC thesaurus*. In the INSPEC database these keywords have also been used, but in addition six other supplementary keywords – *liquid-crystal displays, 7-segment character displays, large-area displays, portable computers, LCDs* and *flat-panel-display market* – taken from the title and abstract are also included. These extra keywords, listed as Free Index Terms, cannot be included in the subject index of the printed version through lack of space. Apart from the inclusion of these

Printed entry in *Electrical & electronics abstracts*

18628   Liquid-crystal displays. D.Pryce.
EDN (USA), vol.34, no.21, p.102-5 108-12, 114 (12 Oct. 1989).
Liquid-crystal displays (LCDs) continue to change the way people view information. From the ubiquitous 7-segment character displays to the large-area displays
used in portable computers, LCDs are making a strong bid to dominate the flat-
panel-display market. (1 ref.)

Computer entry in INSPEC database

000098   Liquid-crystal displays. D.Pryce
EDN (USA), vol 34, no.21, p 102-5 108-12, 114 (12 Oct. 1989) BLDSC. 3661 09400.
Liquid-crystal displays (LCDs) continue to change the way people view information. From the ubiquitous
7-segment character displays to the large-area displays used in portable computers. LCDs are making a
strong bid to dominate the flat-panel-display market. (1 ref.)
Document Type: Journal Paper
Treatment Codes: Application; Practical
Classification Codes: B7260 B4150D
Controlled Index Terms: flat panel displays; liquid crystal displays
Free Index Terms: liquid-crystal displays 7-segment character displays; large-area displays; portable
computers; LCDs; flat-panel-display market.

Fig. 27

keywords, the index of the database will also include all other significant words in the title and abstract. Someone wanting information on changes in methods of viewing information could retrieve this paper from the database by entering the words *view, change* and *information*. In the printed version this would not be possible.

Lastly, it should be noted that more very recent references will be obtained from an online search of a database than could be found in the equivalent printed abstract. This is because databases can be updated with new batches of references much more rapidly than the equivalent batch of references can be produced, bound and distributed in printed form. However, the situation with databases on CD-ROM is slightly different – the costs of publishing a database on CD-ROM are such that it is generally economic to update a disc only every two or three months or so. The references obtained from a CD-ROM search may not therefore be as up to date as those which would be obtained if the equivalent search were carried out on the same database online. This problem can be surmounted, if currency is of vital importance, by first of all searching the CD-ROM version of the database and then going online to retrieve the most recent few months of references.

From what we have said so far it could reasonably be assumed that manual searching is completely obsolete. This is not actually the case for two main reasons. First, most of the databases go back only as far as the late 1960s or early 1970s. If a search has to be made of pre-1970 literature, then it will almost certainly be necessary to use printed abstracts and indexes.

The second and probably most important reason relates to the cost of computerized searching. As we mentioned earlier, online searching generally costs in the region of £1 a minute. Many libraries, although very willing to carry out an online search, do not have the funds to pay for the search themselves. In this situation the cost would have to be met by either the requestor or the department for which he or she works. This problem of payment for an online search is avoidable with CD-ROM databases; as stated previously, once a disc has been bought or leased it can be used continuously without any further payment. However, not all databases are also 'published' on CD-ROM – many are still only available online. In addition, not all libraries can afford to buy them, and those that do often can only afford to buy a handful of the most relevant ones.

Clearly then, the cost of computerized searching can be a deterrent to use though it is important not to take too pessimistic a view. Where a computerized search is wanted and there are no CD-ROM databases either available or relevant, there are ways of keeping the costs as low as possible. Online searching is a very flexible technique and it is possible to restrict the search, and hence the costs, in several ways. One method is to limit the search by date, perhaps to the latest three years, so that only the most recent references are printed out. Another way is to restrict the total number of references produced by excluding papers in foreign languages. Most libraries will have on their staff an intermediary – a person skilled in the techniques of online searching – who will be able to advise on how to achieve an effective result within the cost limits, and also actually to carry out the search itself.

Boolean logic

Although the actual process of searching a database for information is carried out by the computer, it is very useful to have some background knowledge of how the systems work. Online and CD-ROM systems both make use of a type of set theory called Boolean logic, although this may not always be apparent to the user. The way this operates in practice is as follows. A search topic is broken down into its separate parts, known as concepts. For example, a search for information on the possible effect of a vegetarian diet on reducing high blood pressure could be broken down into two main concepts: *vegetarianism* and *blood pressure*. A search for information on the use of remote-sensing techniques for the prediction of earthquakes could be considered to consist of three main concepts: *remote sensing, prediction* and *earthquakes*. A few, though not many, searches consist only of one concept: everything on *neurocomputing*, for example. In a search the computer is instructed to find all the references in the database about each concept, and then to combine these in such a way that the output only consists of those references which include all the concepts; the three Boolean logic operators or links – 'AND', 'OR' and 'NOT' – being used to join the different concepts together.

The combination of concepts using these Boolean operators is usually illustrated by means of Venn diagrams, as in Figure 28. Using 'AND' logic narrows down a search. If the concepts *vegetarianism* and *blood pressure* are combined using 'AND', the shaded section where the two circles intersect will represent the references containing both concepts, that is, *vegetarianism* and *blood pressure*. It can be seen that combining the three concepts *remote sensing, prediction* and *earthquakes* produces an even smaller common set. Using 'OR' logic widens a search. If, in the latter search, the prediction of volcanic activity was also of interest, then the references on *earthquakes* could be combined with those on *volcanoes* to create a larger group or set. 'NOT' logic is used to exclude a particular concept. A search for information on vegetarianism but specifically excluding vegans (i.e. those people whose diet specifically excludes foods of animal origin) would consist of two concepts, *vegetarianism* and *vegans*. As the diagram shows, when these two concepts are combined using 'NOT' logic, a smaller set is produced. It is not used very frequently because there is a danger of missing relevant references. As well as excluding those papers specifically about vegans, those references dealing with vegetarianism in all its aspects, including veganism, would also be lost.

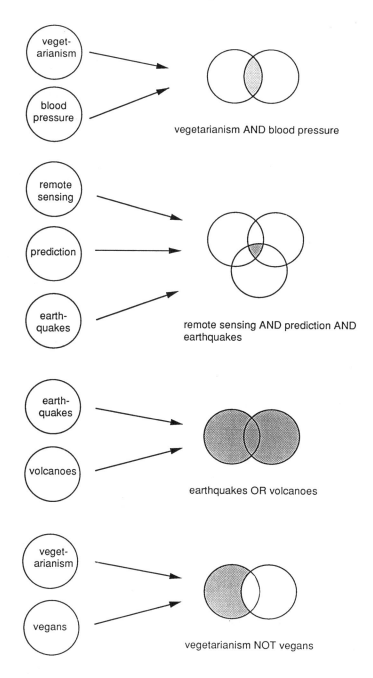

Fig. 28 Examples of Boolean logic

Choosing a database

After analysing the search topic into its component concepts, the next step is to determine which database or databases to use. Often the choice is between several appropriate databases, thus the search for information on use of remote-sensing techniques for earthquake prediction could be carried out on a geological database such as GEOREF, produced by the American Geological Institute, or a more general technological database such as INSPEC or COMPENDEX. Sometimes two or three databases are searched in unison, if it looks as though better results could be obtained by doing so. At other times only one database is really required: information on the corrosion resistance of a certain alloy would probably be most likely satisfied by searching METADEX, the computerized equivalent of *Metals abstracts*.

This choice is further complicated by the fact that many online databases are available via several hosts. INSPEC, for example, can be accessed on at least eight different hosts. MEDLINE is also available on several different online hosts and from at least six different CD-ROM suppliers. Online hosts have alleviated this situation to a certain extent by providing what is in effect an index to all their databases. These work by means of a searcher choosing several likely looking databases, and then typing in the main keywords describing the topic. The host computer then matches these keywords with the databases, displaying the number of occurrences of each keyword in each database – the assumption being that the more times the keyword occurs within a database, the more likelihood there is of the database being relevant. A further advance on this are systems known as 'intelligent gateways' – these are completely separate independent computer systems which have been so programmed that they can use the information entered by a searcher to make a choice between different databases on different host systems. Instead of directly accessing a particular database on an online host, a searcher logs on to the gateway computer via a telecommunications network. The gateway computer, after making the choice of database, then logs on to the chosen host computer.

Given this potentially very confusing variety of systems and databases, the best course of action for any scientist or technologist in any doubt as to which database to search is to discuss the choice with an intermediary who will have the skill and knowledge to advise between different sources.

Choosing keywords

Once a search topic has been broken down into its component concepts, and a database or bases selected, all that remains to do is choose the keywords or subject terms which will describe the concept. Sometimes only one keyword is necessary – *vegetarianism*, for example, has a clearly defined meaning with no obvious synonyms. Frequently, however, a concept may require several keywords to describe it adequately. For example, *high blood pressure* is often referred to as *hypertension*. If the computer is only instructed to search for *blood pressure*, any references which mention *hypertension* but not also *blood pressure* would not be retrieved. Because the computer makes a character-by-character comparison of the keywords input with the keywords in the index, it is also necessary specifically to indicate alternative spellings and word endings. *Earthquakes*, for example, may also be referred to in the singular version as *earthquake*.

If you have some familiarity with the topic, you may be aware of any variant spellings and word endings. Help in determining these is available, however, in the form of the index file of the database. This is an alphabetical list of all the searchable words in the database, and in most if not all computerized information retrieval systems it can be displayed, so that variations in spellings, particularly between British and American practice, and alternative word endings can be checked.

Registry Numbers

Earlier in the book we mentioned the value of using the Chemical Abstracts Service Registry Numbers when searching for information on specific substances. As would be expected, Registry Numbers can be used as a fast reliable way of tracking down relevant references in the *Chemical abstracts* database. This Registry Number system is increasingly being used by other publishers and it is now possible, for example, to use the numbers to search for specific substances in MEDLINE and BIOSIS PREVIEWS (the online version of *Biological abstracts*).

Instructing the computer

A very important aspect of computerized literature searching – and one to which more attention is now being paid – is the mechanism used by a searcher to instruct the computer. For a novice or infrequent user the easiest systems to use are those that rely on menus. No knowledge of the operation of the system is needed; at each point where an action has to be carried out – for example, typing in keywords describing the topic – the menu displays a list of options from which the searcher can choose. With an increasing number of people carrying out their own searches, menu-driven systems are becoming much more widespread – an example of such a system is given on page 63–6.

However, although menu-driven systems are ideal for newcomers, they can be very slow and tedious for experienced searchers, who mainly use command-driven systems. These are systems in which the searcher has to enter commands to instruct the computer what action to carry out. For example, if the searcher wants to print out a set of references, he or she has to enter the appropriate command. Such systems are much faster than those using menus, but of course the searcher does need some prior knowledge of the commands. An example of a command-driven system is given on page 67–8.

Some of the new CD-ROM systems use a combination of menus and commands, intermixed with on-screen prompts. Menus are used at certain points, when, for example, a straight choice has to be made between databases; commands are used for other actions, for example to instruct the system to print out retrieved references. To help inexperienced users prompts or reminders of the most popular commands are displayed on the screen at all times.

Search examples

The effect of a vegetarian diet on high blood pressure

This search was carried out using DIALOG OnDisc™ MEDLINE, a version of MEDLINE supplied by DIALOG^R on CD-ROM. The 'Easy Menu' option, designed for novice searchers, has been used. Note that only a six-month file – January–June 1990 – of MEDLINE was searched. Considerably more references would have been found by repeating the search on the remaining discs, containing earlier years of the database.

In the Easy Menu mode the opening screen (Figure 29a) displays the available search options – in this case the choice was made to search the word/phrase index, thereby instructing the system to retrieve all references in which a chosen word appears. The words *blood pressure* were then typed in, the system responding by displaying the number of records (2,851) in the database which contain this phrase, and in addition listing all the other phrases containing these two words (Figure 29b). After selection, the chosen phrase is displayed accompanied by a menu giving the searcher further choices (Figure 29c). As can be seen from Figure 29d, the option was chosen to modify the existing search – in this case by including the alternative term *hypertension*, giving a total of 5,628 records, containing either the words *blood pressure* or *hypertension* (Figure 29e). The option to modify the search was again chosen, this time by limiting it with the additional concept *vegetarianism*, the result being displayed in Figure 29f (the additional screens showing this process have been omitted). The first of the two records containing the words *blood pressure* or *hypertension* and *vegetarianism* are displayed in Figure 29g.

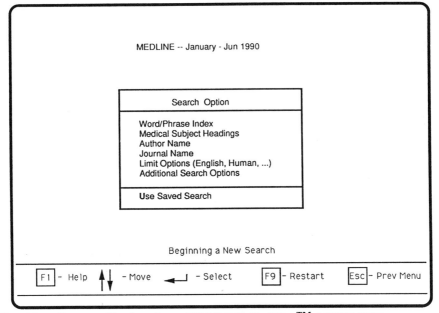

Fig. 29a **Screen display from DIALOG OnDisc™ MEDLINE**

Word/Phrase Index: BLOOD PRESSURE	Records	# Related Terms
BLOOD PRESERVATION --METHODS --MT	25	
BLOOD PRESERVATION --STANDARDS --ST	1	
BLOOD PRESERVATION --VETERINARY --VE	2	F2 for
BLOOD PRESSURE	2,851	5 Related
BLOOD PRESSURE --DRUG EFFECTS --DE	1,499	Terms
BLOOD PRESSURE --GENETICS --GE	4	
BLOOD PRESSURE --PHYSIOLOGY --PH	301	
BLOOD PRESSURE --RADIATION EFFECTS --RE	3	
BLOOD PRESSURE DETERMINATION	151	2
BLOOD PRESSURE DETERMINATION --ECONOMICS --EC	1	
BLOOD PRESSURE DETERMINATION --INSTRUMENTATION --	48	
BLOOD PRESSURE DETERMINATION --METHODS --MT	72	
BLOOD PRESSURE DETERMINATION --NURSING --NU	1	
BLOOD PRESSURE DETERMINATION --STANDARDS --ST	3	
BLOOD PRESSURE DETERMINATION --UTILIZATION --UT	1	
BLOOD PRESSURE MONITORS	57	3
BLOOD PRESSURE MONITORS --ADVERSE EFFECTS --AE	2	
BLOOD PRESSURE MONITORS --STANDARDS --ST	3	
BLOOD PRESSURE, HIGH		0 Entries Selected
BLOOD PRESSURE, LOW		0 Records Found
BLOOD PRESSURE, VENOUS		

F1 - Help   ↑↓ - Move   ◄┘ - Select   F10 - When Done   Esc - Prev Menu

Fig. 29b **Screen display from DIALOG OnDisc™ MEDLINE**

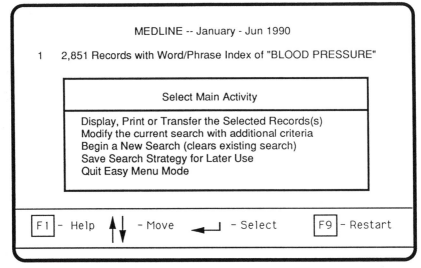

Fig. 29c Screen display from DIALOG OnDisc™ MEDLINE

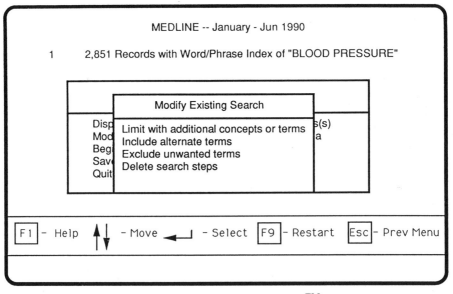

Fig. 29d Screen display from DIALOG OnDisc™ MEDLINE

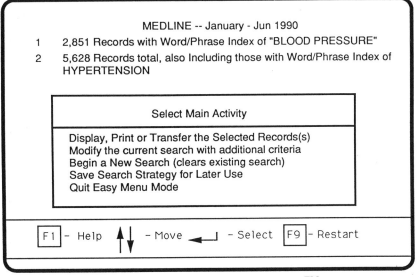

Fig. 29e   Screen display from DIALOG OnDisc™ MEDLINE

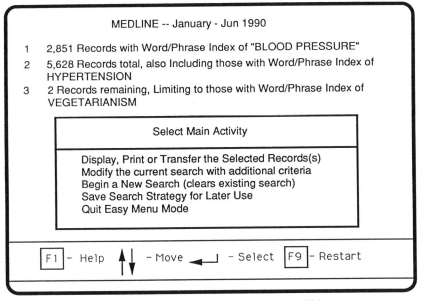

Fig. 29f   Screen display from DIALOG OnDisc™ MEDLINE

```
  1 of 2  Complete Record

  90099157
  Lifestyle intervention in hypertension.
  Marley J
  Practitioner  (ENGLAND)  May 8 1989  233 (15680 p661-3  ISSN: 0032-6518
  Language: ENGLISH
  Journal Announcement: 9004
  Subfile: INDEX MEDICUS
  Patients are often interested in taking some responsibility for treating
  their own illnesses, especially in a symptom-free condition such as
  hypertension. Confusing information is available on non-pharmacological
  measures, which often results in none of these treatments being offered to the
  patient. This review discusses what is of value, what may be helpful and what
  is likely to be worthless.
  Tags: Comparative Study; Human
  Descriptors: *Hypertension--Therapy--TH; *Life Style; Acupuncture Therapy;
  Diet, Reducing; Diet, Sodium-Restricted; ,Dietary Fats--Administration and
  Dosage--AD; Exercise; Temperance; Trace Elements--Administration and Dosage--AD;
  Vegetarianism


           ***  End of Record  ***    Next: Ctrl+PgDn


  F1-Help  F4-Format  F5-Sort  F7-Records  F8-Print/Xfer  Esc-Main Menu
```

Fig. 29g   Screen display from DIALOG OnDisc™ MEDLINE

The use of remote-sensing techniques for the prediction of earthquakes
This search (shown in Figures 30a and 30b) was carried out on the
COMPENDEX*PLUS database, using DATASTAR, a Swiss-based host
accessed by a telecommunications link. The logging-in procedures, connect-
ing the searcher's terminal to the system's computer, have not been shown.
DATASTAR, although a command-driven system, prompts the searcher for
input – to distinguish between the searcher's input and the computer's
prompts and responses the former have been underlined. One feature
illustrated in this search is the use of phrases or compound words – terms
such as *remote sensing*. It is possible to instruct the computer to search for
two words next to each other, usually by inserting a symbol known as a
proximity operator, to tell the computer that the two words must be adjacent.
In the DATASTAR system the symbol ADJ is used, so that *remote sensing*
is entered as *remote* ADJ *sensing*; other systems will use a different symbol
but the end result will be the same.

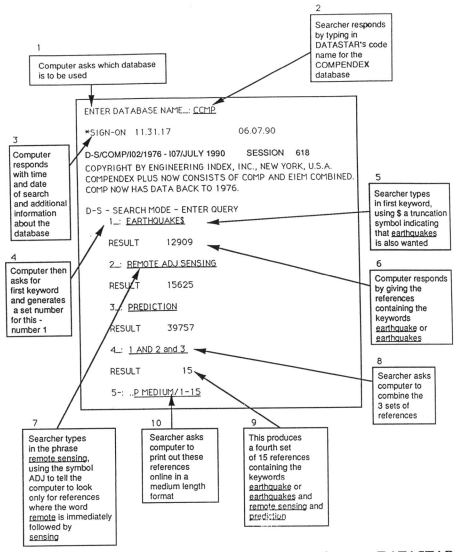

1
Computer asks which database is to be used

2
Searcher responds by typing in DATASTAR's code name for the COMPENDEX database

3
Computer responds with time and date of search and additional information about the database

4
Computer then asks for first keyword and generates a set number for this - number 1

5
Searcher types in first keyword, using $ a truncation symbol indicating that earthquakes is also wanted

6
Computer responds by giving the references containing the keywords earthquake or earthquakes

7
Searcher types in the phrase remote sensing, using the symbol ADJ to tell the computer to look only for references where the word remote is immediately followed by sensing

8
Searcher asks computer to combine the 3 sets of references

9
This produces a fourth set of 15 references containing the keywords earthquake or earthquakes and remote sensing and prediction

10
Searcher asks computer to print out these references online in a medium length format

```
ENTER DATABASE NAME_: CCMP

*SIGN-ON  11.31.17              06.07.90

D-S/COMP/I02/1976 - I07/JULY 1990    SESSION  618
COPYRIGHT BY ENGINEERING INDEX, INC., NEW YORK, U.S.A.
COMPENDEX PLUS NOW CONSISTS OF COMP AND EIEM COMBINED.
COMP NOW HAS DATA BACK TO 1976.

D-S - SEARCH MODE - ENTER QUERY
     1_: EARTHQUAKE$

     RESULT     12909

     2_: REMOTE ADJ SENSING

     RESULT     15625

     3_: PREDICTION

     RESULT     39757

     4_: 1 AND 2 and 3

     RESULT       15

     5-: ..P MEDIUM/1-15
```

Fig. 30a   Print-out from COMPENDEX*PLUS database on DATASTAR

These two simple searches were chosen to demonstrate the basic principles of computerized searching: the creation of two or more sets of references and the combination of these into a final set, consisting of a manageable number of relevant references. Not all searches, of course, work out in

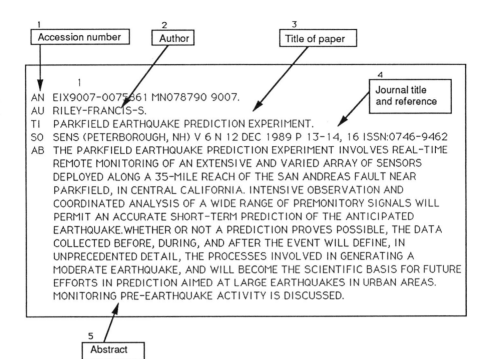

Fig. 30b Print-out from COMPENDEX*PLUS database on DATASTAR

practice as simply as these. It is frequently necessary to change the strategy of a search while in progress. Sometimes the combination of concepts produces almost no references and then it is necessary to drop the least important concept or widen the scope of a concept by introducing more keywords. Sometimes the reverse situation applies, and too many references are produced. It may then be necessary to restrict the scope of the search by introducing further concepts or to reduce the number in some other way – perhaps by limiting the search to references less than three years old or by excluding foreign language material. There are also other more sophisticated techniques which can be used to decrease the number of references obtained and at the same time to increase the relevancy.

There is, in fact, an inverse relationship between the relevancy and the total number of references which are produced from a search. If the relevancy is increased then this will be at the expense of the number of references retrieved. Not every reference produced in a computerized search is likely to be relevant. It is often cheaper in an online search to accept a slightly larger number of references, some of which may be of marginal interest, than to spend a longer time online trying to cut down the size of the output.

Full-text databases

Not all databases are bibliographical in content – that is, consist solely of references to journals, conference papers and so on. An increasing number now consist of the complete or full text of publications such as journals, encyclopedias and reference books. For example, both the Royal Society of Chemistry and the American Chemical Society's major journals can be accessed online; and the *McGraw-Hill encyclopedia of science & technology*, illustrated in Chapter 2, has been 'published' on CD-ROM. There are many more examples – what they all have in common is the fact that the whole text of an article, not just the reference and summary, can be searched and displayed on a VDU. The big advantage of these databases is that the information is immediately available – no more waiting for a paper that may or may not be relevant. However, because the complete text of articles is searched, there is a bigger likelihood of retrieving items of fringe interest in which the required information is given only a brief mention.

Numeric databases

A third category of database – that giving numerical and factual data on the structure, nomenclature or physical and chemical properties of substances – has also emerged. Like their bibliographical and full-text counterparts, many of these databases can be searched online or on CD-ROM – a few are available using both methods. The sample record shown in Figure 31 gives some indication of the kind of information which these types of database contain. Some include references to the original sources of information, others do not; the content of these databases tends to be more variable than bibliographical and full-text ones. A few also have facilities to carry out calculations on the data obtained.

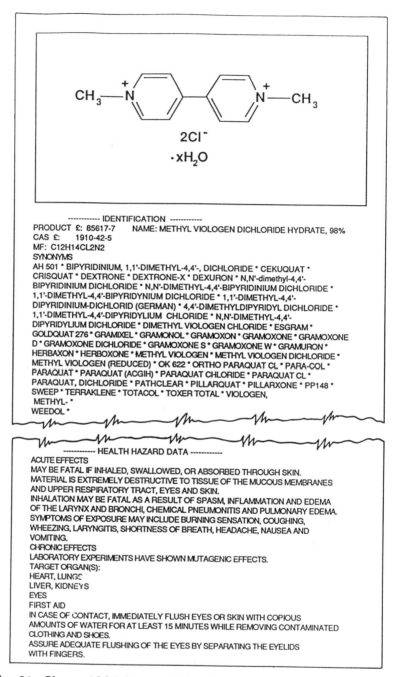

------------ IDENTIFICATION ------------
PRODUCT £: 85617-7   NAME: METHYL VIOLOGEN DICHLORIDE HYDRATE, 98%
CAS £:   1910-42-5
MF: C12H14CL2N2
SYNONYMS
AH 501 * BIPYRIDINIUM, 1,1'-DIMETHYL-4,4'-, DICHLORIDE * CEKUQUAT *
CRISQUAT * DEXTRONE * DEXTRONE-X * DEXURON * N,N'-dimethyl-4,4'-
BIPYRIDINIUM DICHLORIDE * N,N'-DIMETHYL-4,4'-BIPYRIDINIUM DICHLORIDE *
1,1'-DIMETHYL-4,4'-BIPYRIDYNIUM DICHLORIDE * 1,1'-DIMETHYL-4,4'-
DIPYRIDINIUM-DICHLORID (GERMAN) * 4,4'-DIMETHYLDIPYRIDYL DICHLORIDE *
1,1'-DIMETHYL-4,4'-DIPYRIDYLIUM CHLORIDE * N,N'-DIMETHYL-4,4'-
DIPYRIDYLIUM DICHLORIDE * DIMETHYL VIOLOGEN CHLORIDE * ESGRAM *
GOLDQUAT 276 * GRAMIXEL * GRAMONOL * GRAMOXON * GRAMOXONE * GRAMOXONE
D * GRAMOXONE DICHLORIDE * GRAMOXONE S * GRAMOXONE W * GRAMURON *
HERBAXON * HERBOXONE * METHYL VIOLOGEN * METHYL VIOLOGEN DICHLORIDE *
METHYL VIOLOGEN (REDUCED) * OK 622 * ORTHO PARAQUAT CL * PARA-COL *
PARAQUAT * PARAQUAT (ACGIH) * PARAQUAT CHLORIDE * PARAQUAT CL *
PARAQUAT, DICHLORIDE * PATHCLEAR * PILLARQUAT * PILLARXONE * PP148 *
SWEEP * TERRAKLENE * TOTACOL * TOXER TOTAL * VIOLOGEN,
 METHYL- *
WEEDOL *

------------ HEALTH HAZARD DATA ------------
ACUTE EFFECTS
MAY BE FATAL IF INHALED, SWALLOWED, OR ABSORBED THROUGH SKIN.
MATERIAL IS EXTREMELY DESTRUCTIVE TO TISSUE OF THE MUCOUS MEMBRANES
AND UPPER RESPIRATORY TRACT, EYES AND SKIN.
INHALATION MAY BE FATAL AS A RESULT OF SPASM, INFLAMMATION AND EDEMA
OF THE LARYNX AND BRONCHI, CHEMICAL PNEUMONITIS AND PULMONARY EDEMA.
SYMPTOMS OF EXPOSURE MAY INCLUDE BURNING SENSATION, COUGHING,
WHEEZING, LARYNGITIS, SHORTNESS OF BREATH, HEADACHE, NAUSEA AND
VOMITING.
CHRONIC EFFECTS
LABORATORY EXPERIMENTS HAVE SHOWN MUTAGENIC EFFECTS.
TARGET ORGAN(S):
HEART, LUNGS
LIVER, KIDNEYS
EYES
FIRST AID
IN CASE OF CONTACT, IMMEDIATELY FLUSH EYES OR SKIN WITH COPIOUS
AMOUNTS OF WATER FOR AT LEAST 15 MINUTES WHILE REMOVING CONTAMINATED
CLOTHING AND SHOES.
ASSURE ADEQUATE FLUSHING OF THE EYES BY SEPARATING THE EYELIDS
WITH FINGERS.

Fig. 31 Sigma-Aldrich material safety data sheet on CD-ROM

Summary

● Find out what online and CD-ROM searching facilities are available in your library or information centre. Ask if the cost of an online search will be met by the library or whether you will have to make your own arrangements for payment.

● Before starting a search, determine its subject content as specifically as possible and include any synonyms known to you.

● Try to be present if an online search is carried out for you. Don't talk too much: it will slow down the search, which will therefore cost more.

● Be wary of any search which produces absolutely no references on a topic. Misspelling a keyword or incorrect use of a command will give a null result.

5

Obtaining and organizing information

Obtaining references

By this stage, whether your literature search was carried out online, on CD-ROM or using printed abstracts and indexes, you will almost certainly have found some relevant references: maybe only four or five but possibly fifty or more. Whatever the number, it is clearly important to obtain a copy of the original publication as soon as possible. Although colleagues might be able to supply you with the occasional item, in most instances you will have to rely on the resources of your library or information centre. The obvious first step is to check to see if the publications you need have been bought by the library and are available on-site. People familiar with a particular library often forget to do this as it is very easy to assume that any relevant material would have been noticed while browsing round the shelves. However, you will probably discover that at least one of the references you need is not in stock – rapidly escalating book and journal prices have meant that even large university libraries are no longer self-sufficient – and will have to be obtained from outside the organization.

Interlibrary loan

Although libraries will consider buying publications needed by readers, it is generally faster to obtain the item (or a photocopy of it) from another library, a procedure known as interlibrary loan or ILL. British scientists and technologists are fortunate in that a sophisticated interlibrary loan service has been developed in the UK. Based at the British Library's warehouse at Boston Spa in Yorkshire is a very large collection of scientific and technical journals and books, making a fast comprehensive service for loans possible. You will almost certainly be asked to fill in a form or card, such as the one in Figure 32, for each publication you want. Note that the form asks for the source of the reference. This is a common precaution against inaccuracies: if the reference is wrong it may still be possible to track down the correct publication from the original source.

```
INTER LIBRARY LOAN JOURNAL REQUEST                          Date

Title of Journal in full                                   For Library Staff Use


Year        Volume      Part        Pages


Author of Article


Title of Article


Your Borrower Number        Your Department        Where did you find the
                                                   details for this request?
P

This request will be of no use to me after:

PLEASE WRITE YOUR NAME AND ADDRESS OVERLEAF

This request is now available for collection. Please bring
this card to the Library before:
```

Fig. 32 Interlibrary loan request card

One aspect of an interlibrary loan service which ought to be mentioned at this point is the cost. The service is not free: the cost of obtaining a single loan from the British Library in 1991 was nearly £3.75. Although most libraries do not pass on the charges to their clientele, limited budgets have caused many to attempt to restrict demand in some way, often by setting an upper limit on the number of requests which can be made at any one time. While such restrictions are necessary evils, they tend to create a negative impression amongst readers. No library would want you to be deterred from requesting items important for your work, so do ask if you need help.

Offprints

In addition to the interlibrary loan service, another way of getting hold of papers is to ask the author for an offprint. On publication of a paper most journals provide authors with a number of free copies or offprints (sometimes called reprints) to distribute themselves. Some of these are immediately given to colleagues and friends, the remainder are sent out to anyone expressing an interest in the work. Provided a request is made soon after publication, the chances of obtaining an offprint in this way are high. Many research teams have their own request card like the one in Figure 33. Requesting offprints directly from authors, although sometimes slow, is particularly useful where a paper contains photographs which do not copy well. It also tends to be used by scientists and technologists as an informal way of signalling their interest in a piece of research.

```
┌─────────────────────────────────────────────────────┐
│                                                      │
│                                                      │
│        Microbiology Research Group,                  │
│        Department of Pharmaceutical Sciences,        │
│        Aston University,                             │
│        Aston Triangle,                               │
│        Birmingham B4 7ET.                            │
│        United Kingdom.                               │
│                                                      │
│                                                      │
│        We would very much appreciate a reprint       │
│        of your paper which appeared in               │
│                                                      │
│                                                      │
│                                                      │
│                                                      │
│        Thank you,                                    │
│                                                      │
│        Yours sincerely,                              │
│                                                      │
│         This part of the card may be detached for reply │
├─────────────────────────────────────────────────────┤
│                                                      │
│        Microbiology Research Group,                  │
│        Department of Pharmaceutical Sciences,        │
│        Aston University,                             │
│        Aston Triangle,                               │
│        Birmingham, B4 7ET                            │
│        United Kingdom                                │
│                                                      │
│                                                      │
└─────────────────────────────────────────────────────┘
```

Fig. 33   Offprint request card

Online document ordering

References found during an online search can often be ordered at the time of the search – database producers and other organizations, such as the British Library, being used to supply the publications. You will probably find, though, that the intermediary carrying out your search will not make use of this option, partly because there is no opportunity to check if the references are already available on-site, and partly because it is necessary to make an immediate decision on the relevance of each item.

Urgent information

Using the interlibrary loan service is a reliable, convenient way of getting hold of publications, but there will be a wait of a few days at the very least before the material arrives. For the majority of requests this delay, though not ideal, is generally acceptable. If you want a paper, book or other publication urgently, you will have to try other methods of supply. The British Library operates an Urgent Action Service; although more expensive than the normal service, this guarantees that if a specially requested item is in stock, it will be sent off the same day by first-class mail.

Another fast method of document supply is to use fax machines to transmit information between libraries. Using fax to send copies of documents makes it possible for a library to provide a same-day service, but there are some difficulties. Not all libraries own or rent a machine, the transmission costs are quite high and legibility is not always 100%. This is not to say that you should be deterred from asking for fax copies, only that you should be aware of the snags involved.

Another option is to ask the staff in your library or information centre if they could find out whether the item is in stock at another library nearby. Other libraries will not lend directly to non-members, but you should be able to use the item on the premises.

Copyright legislation

This outline would not be complete without mentioning copyright legislation. This aims to keep a balance between the rights of authors and publishers to obtain a fair return for their work and the right of the public to have access to knowledge. Probably the aspect of copyright legislation of most interest to scientists and technologists is that concerning the photocopying of papers from journals. In the UK, the 1956 Copyright Act allowed single copies of articles required for research or private study to be made from journals without payment of royalties. This position would appear to be unaffected by new legislation, the Copyright Designs and Patents Act, passed in 1988. However, the new Act, when fully introduced, will not permit 'multiple' copying of articles without payment although how 'multiple' is to be defined has not yet been clarified. Another aspect of the new legislation which may affect scientists and technologists is that relating to downloading. Downloading is the transfer and storage of references from a database to a microcomputer – its applications are outlined in the next section. The 1988 Act does not specifically mention downloading (and it is not yet entirely clear how the Act will be interpreted in this area). In practice, however, an agreement or 'licence' needs to be obtained from the owner of the database before downloading takes place. In the USA a copyright law introduced in 1978 allows a maximum of five articles in any one year to be copied from the issues of any one journal published over the previous five years. Royalty payments are required for any photocopying made outside these provisions.

Organizing personal collections

Most scientists and technologists will have experienced the problem of trying to find a particular reference or piece of information they know to be in their own personal literature collection and the subsequent irritation if the search proves fruitless. It is obviously impossible to set exact limits, but above a certain size an individual collection kept in no particular order will tend to become unwieldy. Added to this is the fact that most individual scientists' literature collections are really mixtures of the complete texts of papers in the form of reprints and photocopies, and references to some material not actually in their possession: books, conferences and the like. Many of these references are likely to be on odd scraps of paper, backs of envelopes, cash till receipts and so on.

Laying down hard and fast rules about how to organize a literature collection would obviously be inappropriate – people with photographic memories may be quite happy with a few hundred papers in no particular order, whereas those who like order and tidiness could probably not tolerate a pile of photocopies, reports and scraps of paper heaped up in the corner of an office. What we can do is to give some general guidelines and point to some current developments which are likely to make the task easier.

The simplest solution is to depend on the physical organization of the material alone, arranging papers according to fairly specific subject groups. Information for which you have the reference only but no original would have to be filed separately, preferably using the same subject groups. Having more than one place to look can be a nuisance: it is easy to overlook one of them, and you would have to rely on memory to find papers by particular authors. The system does have the great advantage, though, of needing little time or effort to set up and maintain; organizing and indexing a personal collection will inevitably be low down on a list of priorities when time is short. Another simple alternative is to write each reference on an ordinary index card, and store these in author name order. Coloured labels or symbols can then be added to each card to indicate the subject fields, as in Figure 34.

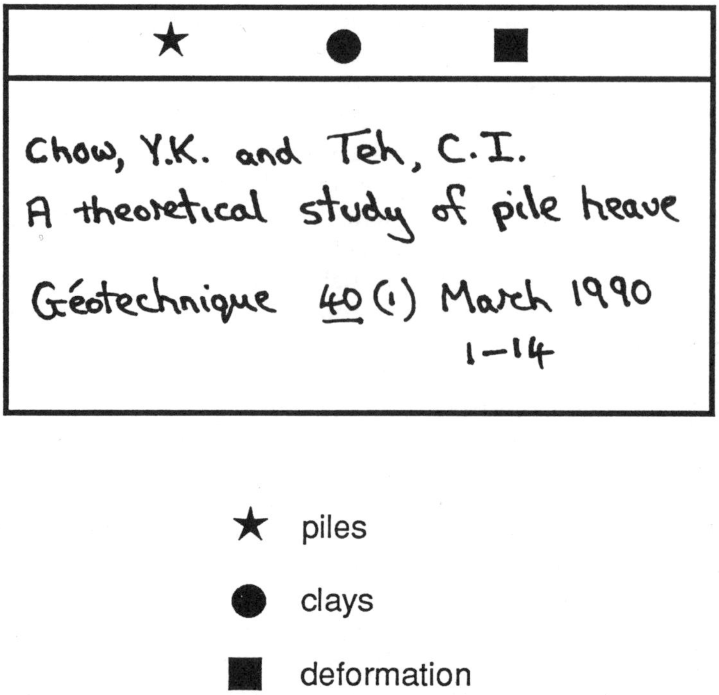

★   piles

●   clays

■   deformation

Fig. 34 Reference in a manual indexing system

When references on a particular topic, such as *deformation*, are required, these can be found by extracting all cards containing the appropriate label, in this case a square.

These types of systems are clearly best suited to fairly small personal collections, used by one individual. Collections which are likely to reach a reasonably large size, particularly those which are shared between a group of people, really need some kind of subject and author indexing. Until a few years ago indexing had to be carried out by manual means, but the introduction of cheap microcomputers has made it possible for individuals and groups to set up their own personal databases.

Computerization

A computerized personal literature collection or database can offer considerable advantages over a manual collection. However, it is important not to overestimate what a microcomputer can do. To establish and maintain a reasonable database requires planning and effort and it is best to be clear from the outset what the advantages and disadvantages will be. There are two main benefits. First, computerization, if carried out effectively, will give much improved searching facilities compared with a manual system. This is because a computerized system can search all the separate parts of a reference, including all the words in the title and any keywords or indexing terms added to the reference by the individual setting up the database. In a computerized system this sort of search can take seconds; in a manual one it could take several hours.

Secondly, computerization can provide more efficient management of a literature collection. With a suitable software package, it is now possible to make a computerized search of a database (either on CD-ROM or using an online host such as DATASTAR), download the references and incorporate them directly into a personal database. As more and more information becomes available in electronic form this facility will become increasingly important. In addition, a computerized system can provide advantages when preparing a paper for publication. Journals usually require references to be in a certain layout or format. Some, for example, require the author's name to be followed by the year of publication, as in the reference below.

Chow, Y. K. and Teh, C. I. (1990) A theoretical study of pile heave. *Géotechnique 40* 1–14.

Others specify that the author's name is to be followed by the title of the paper, then the journal details:

Chow, Y. K. and Teh, C. I. A theoretical study of pile heave. *Géotechnique 40* (1) March 1990 1–14.

In a manual system this reformatting has to be done by the author. In a computerized system (again with the proviso of an appropriate software package) it is possible to extract references from a database and have these references formatted by the package into the required layout.

So much for the benefits. What are the problems? First, software: there are a large number of software packages that could be used to set up a personal database. The difficulty is that many of them do not offer all the requisite features at a reasonable price. In addition the software is designed for the storage and retrieval of text and numbers, and current packages generally do not have the ability to record structural information about compounds. Secondly, a significant amount of time will be taken up by entering references into the system. Unless you are a skilled keyboarder, the input time will not be very different from that required to record references on cards in a manual system. For a poor keyboarder it could take quite a lot longer. Thirdly, time has to be allocated for familiarization with both the package and the set-up procedures associated with it. Some people consider that there is a fourth disadvantage of a computerized system, namely the possibility of the database being accidentally destroyed, but good back-up procedures will obviously prevent this.

Before making a decision on whether to set up a personal database, there are a number of practical considerations which are worth bearing in mind.

- Access to a microcomputer. If a database is to be regularly used and updated it must be easily accessible. The optimum solution is for it to be located within an individual's own office or within a shared laboratory for a research group.

- A hard disc. This is virtually a necessity. A floppy disc is unlikely to be able to accommodate more than 500 or so references, and it would obviously be very inconvenient to have to search separately through two or more floppy discs for relevant items.

- Whether a new database is to be created or an existing one is to be converted into a computerized form. It is obviously better to set up a database at the start of a research project rather than wait a couple of years when possibly 300 or 400 references have already been recorded in a manual format on cards.

- Whether a database is to be used by an individual or shared between several colleagues. A shared database will obviously be larger than that belonging to an individual and more complex searches are likely to be carried out. A computerized literature collection scores heavily over a manual one where good search facilities are needed. In addition, in an individual manual system memory frequently plays an important part in the retrieval of references but a shared system has to rely on the adequacy of the indexing and searching facilities.

Personal database software
If a decision is made to set up a database, then the next step is to find a suitable software package. There are now several packages around which could be used for this purpose. Some, for example Pro-Cite (Figure 35) or Reference Manager, are intended specifically for the management of personal literature collections. Others, such as Cardbox, are general database packages. Whatever the type of package, however, they all have to be able to carry out certain basic operations. These operations fall into the following categories.

```
┌─────────────────────────────────────────────────────────────────────┐
│                                                                       │
│     Pro-Cite Sample 1.4                                               │
│     Database: A:\SAMPLE                    26 Records   26 Selected    │
│     ─────────────────────────────────────────────────────────────     │
│                                                                       │
│     ┌──────────┐                                                      │
│     │   Edit   │   Edit, Insert and Delete Records                    │
│     ├──────────┤                                                      │
│     │  Select  │   Select Records: Searching, Duplicates, In-text References │
│     ├──────────┤                                                      │
│     │   soRt   │   Sort Selected Records                              │
│     ├──────────┤                                                      │
│     │  Print   │   Print Selected Records to Screen                   │
│     ├──────────┤                                                      │
│     │  Merge   │   Merge Selected Records                             │
│     ├──────────┤                                                      │
│     │  Delete  │   Delete Selected Records                            │
│     ├──────────┤                                                      │
│     │  Index   │   Index Selected Records                             │
│     ├──────────┤                                                      │
│     │ Options  │   Change Layout, Renumber Records, Modify Workforms  │
│     ├──────────┤                                                      │
│     │  Change  │   Change Database                                    │
│     ├──────────┤                                                      │
│     │   Quit   │   Close Database and Return to DOS                   │
│     └──────────┘                                                      │
│                                                                       │
└─────────────────────────────────────────────────────────────────────┘
```

Fig. 35 Main menu screen display from Pro-Cite

Creating a record format

Before a reference can be entered into a computer, a suitable record format or data structure must exist. A record format for a reference would normally consist of a number of separate parts or fields. Each field would store a part of the reference, so, for example, there would be an author field, a title field, possibly an abstract field and so on. With some packages, such as Cardbox, users set up their own record formats, making their own decisions about which fields to include, how long they should be and what they should be called. Packages intended specifically for literature collections often provide pre-defined record formats, called workforms or templates. Pro-Cite, for example, has 20 different workforms, covering different material types such as books, journal articles and theses.

Pro-Cite and Cardbox are both examples of software packages which utilize variable length fields. This means that the length of the field can vary from record to record depending on the information contained within the field. A title such as *'Non-linear analysis of laterally loaded piles in heavily overcon-solidated clays'*, for example, will require a much longer field than the title in the reference mentioned previously, *'A theoretical study of pile heave'*.

81

Alternatively, fields can be of fixed length, which means they always consist of the same number of characters and spaces regardless of their contents. For example, a fixed length field for a title might be 60 characters in length. This would be sufficient to store the second title but not the first. Variable length fields are preferable for literature collections because of the wide variation in length of references, but not all packages provide such a feature.

Adding new records

The simplest, most time-consuming method is direct entry; that is, typing in the references on the keyboard. In addition to this process, however, some packages can transfer or import records from other databases – a useful time-saving feature. The suppliers of the personal database type software have tended to establish special modules or programs which will reformat or convert references from online and CD-ROM databases into the appropriate format for the personal database. A different program is required for each online host, however, so these tend to be limited to the larger ones, such as DIALOG^R. Whether or not this facility exists, a package should certainly be capable of both accepting and creating records in ASCII format. This is a standard computer method of storing text and numbers and will make it possible to transfer the references in the database from one package to another. It therefore keeps open the option of upgrading to another software package – without it the whole of the database would have to be rekeyed.

Indexing

With some software packages, such as Cardbox, keywords or indexing terms can be taken from the various parts of the reference – for example, title and abstract – and used to form a separate index (known as an inverted index). The advantage of a separate index is that searching the database is faster. The disadvantage is that the index takes up considerable storage space on the hard disc. If there is no separate index each reference in a database has to be searched in turn, this being known as serial searching. Serial searching is too slow for large databases but can be perfectly adequate for small databases, the size of a personal literature collection.

Whether or not the software is capable of creating a separate index, someone setting up a database will need to decide if it is worth adding extra keywords or indexing terms to each reference. The paper mentioned previously ('A theoretical study of pile heave') is concerned with deformation in soil, although this is not specifically mentioned in the title. If only the title and journal details are entered into the database, this reference will not be retrieved when the computer is instructed to look for deformation. Often a researcher is interested in a paper only because a particular method has been used; this is also very unlikely to be mentioned in the title. These types of problems can be avoided by the inclusion of keywords or indexing terms for each reference. These keywords are stored in a separate field of the record, as in the example in Figure 36 from a database under the Bib/SEARCH text retrieval package. If an abstract is also included as part of a reference, then the need to include indexing terms may not be quite so great. This is because even if the title does not mention a certain topic, it could appear in the abstract. However, there can be no certainty that an abstract will cover all aspects of a paper. The only way to be sure of this is to include these aspects as indexing terms.

Some practical points to note for anyone adding indexing terms to references are:

- The indexing terms used to describe a reference should be as specific as possible. Indexing references generally rather than specifically can mean retrieving many items at each search. If items are indexed exactly it is easy if nothing is found using a specific word to broaden the search to more general words or phrases.

```
1Help    2Colors 3Check  4Assist 5Control6Next       9Last 10Quit
 FIELD   SET 2[1:1]          |  REC [9:9]FILE:LAMBERT
AUTHOR  |Chow, Y.K.
AUTHOR  |Teh, C.I.
TITLE   |A theoretical study of pile heave
DATE    |1990
PAGE    |1-14
LOCATN  |Filing Cabinet
JOURNAL |Geotechnique
VOLUME  |40
ISSUE   |1
KEYWORD |analysis
KEYWORD |clays
KEYWORD |deformation
KEYWORD |elasticity
KEYWORD |piles
KEYWORD |soil-structure interaction
TEXT    |Theoretical study of vertical soil movement and pile
        |heave due to installation of a driven pile in clay

Ctrl-I (Index)  Ctrl-D (delete field)              INSERT
```

Fig. 36   Screen display from Bib/SEARCH

- It is important to be as consistent as possible when choosing indexing terms; for example, always using the term *clays* rather than sometimes *clays* and sometimes *clay*. The way authors cite their names can also vary, for example *P. A. Lambert* or *Peter A. Lambert*; if one form of name is chosen and kept to, it can make searching much easier. Some software packages, such as Pro-Cite, will enable users to set up authority lists – these are lists of all the words which have previously been used as indexing terms. With another software package, Reference Manager, it is possible to have such a list constructed automatically.

Searching

Most packages will allow all fields of a record to be searched. This means that if the word being searched for appears anywhere within the title of the paper, journal or conference title, indexing terms or abstract (if included), the reference will be retrieved. With some packages it is possible to specify which field is to be searched, for example only the title field. Restricting a search to a specific field in this way can be useful if too many references are found in a search of the whole database.

Two other common software features often found in this type of software are Boolean searching and truncation. Boolean searching is based on the principle of Boolean logic, described in Chapter 4. This allows search words to be combined together. The computer could be instructed, for example, to look for the search words *deformation* and *piles*. It would then retrieve any references containing both these search words. Truncation, also mentioned in Chapter 4, is a method of allowing a computer to search for a word stem or part of a word. By far the most common form is right-hand truncation in which the ending of a word is cut off; for example, *heav*. The computer will then retrieve all references containing this stem, such as *heave* and *heaving*. Many of the software packages for personal databases do offer right-hand truncation and it is a very useful feature.

Output

In a personal database there will be a need for two different types of output. First is the display and printing out of a few references retrieved from the database for personal use. Since the format or layout will be of no particular importance these can be printed out directly.

The second type of output is a list of references to accompany the publication of a paper. As mentioned previously, the editors of journals usually require references to be set out or formatted to a particular style. Unfortunately since there is little standardization of practice there are numerous different styles. The personal database type software packages have tried to meet this need. Pro-Cite, for example, provides over 20 different output format styles, such as those required for the *Journal of the Chemical*

Society and *Physical review*. With the wide variation in styles, however, it would be impossible for a package to cover all those possible. They therefore also provide the facility for someone to define their own format or layout. Some packages such as Reference Manager and Bib/SEARCH are also able to search through a manuscript for references to papers and then compile a list of all these, so saving the author the job of extracting these from the database. A further feature generally available is the ability to sort references awaiting output. This can be either by author name or by publication year and can be useful in the preparation of a manuscript for publication.

Software packages

There are, as stated earlier, a fairly large number of software packages which could be used to set up a personal database management system. Many of these, however, are likely to prove both too expensive and too sophisticated for the average student or researcher. The ones listed below are (at the time of writing) amongst the main packages in use for both individual and group literature collections. More software can be expected in the future, as the requirement for personal databases increases.

askSam
askSam Systems, PO Box 1428 Perry, FL 32347, USA

BiB/SEARCH
Information Automation Ltd, Penbryn, Bronant, Aberystwyth, Dyfed SY23 4TJ

Cardbox Plus
Business Simulations Ltd, 30 St James's Street, London SW1A 1HB

Get-A-Ref™
DatAid™ Inc., P.O. Box 8865, Madison, WI 53708−8865, USA

ideaList
Blackwell Scientific Software, Blackwell Scientific Publications, Osney Mead, Oxford OX2 0EL

Papyrus
Research Software Design, 2718 SW Kelly St, Suite 181, Portland, Oregon 97201, USA

Pro-Cite
Personal Bibliographic Software, Inc., Woodside, Hinksey Hill, Oxford OX1 5AU

Reference Manager
Research Information Systems, Inc., 1991 Village Park Way, Suite 205, Encinitas, CA 92024, USA

Summary

- Use the interlibrary loan service to get hold of references which are not available locally. Alternatively, write to the authors of recently published journal papers, asking for an offprint.

- If the references are needed urgently, ask the library or information centre staff if there are any special services such as fax available.

- If not, check to see if you could use the item in another local library.

- Give some thought at the beginning of your project as to how your literature collection would best be organized.

6

Keeping up to date

Having found and evaluated the literature in your subject field, the next step is to make sure you keep up to date with current information and development. You do have certain advantages in this respect: you should know the field well enough to be able to be fairly selective in what you read, and you should also have a good idea of which individuals and journals are likely to publish relevant information. The main difficulty you are likely to face will be lack of time to read all the information available: the actual process of finding the information is relatively simple.

There are really two different ways of keeping up to date. One is to go to meetings at which new techniques and developments are being discussed, and to supplement these with news and tips from colleagues; obviously most of the information picked up in this way will be verbal, rather than written, in form. The second way is to scan relevant journals, lists of new books and so on – that is, make use of published information. In practice, of course, these two methods complement one another, with most people using a mixture of both.

Verbal information

Most scientists and technologists are delighted to discuss the latest developments in their fields with interested colleagues. You can save yourself a great deal of time and effort in searching the literature by discussing your problem with someone else. It is worth contacting a person who, for example, might have published recently in your field of interest. The informal atmosphere at a conference provides an ideal opportunity for what could well prove to be a valuable discussion. You might be made aware of a paper or technique your literature search had missed, perhaps in a separate area of research, which could be applied to your own problem. You might also learn about things to avoid, such as experiments which failed and will never be published. Even allowing for some understandable reluctance to release information not yet published, you can often get ahead of the literature by learning of a forthcoming paper 'in press'.

Published information

Information picked up at conferences and from colleagues and so on is very useful but it obviously cannot be depended upon totally. It might be inaccurate – there will have been no screening process to compare with the refereeing that takes place when papers are submitted to journals. Since the information is not acquired systematically, it would be very easy to miss something important. It is necessary, therefore, to supplement this verbally acquired knowledge with newly published information.

Publishers' advertising

A very easy way of finding out about new books is to write to as many scientific, technical and medical publishers as possible, stating your subject interests and asking to be added to their mailing lists. Publishers put a lot of effort into producing promotional literature. It can be difficult for them to target it at the right audience, and most will be pleased to send you information regularly about new and forthcoming publications (Figure 37).

Fig. 37 Elsevier mailing list cards

Some of the larger booksellers also publish their own catalogues and send them out on a regular basis.

Scanning journals

One of the commonest ways of keeping up to date with new papers is to scan through every new issue of interesting and relevant journals. It will not be possible, of course, to look at all the journals which sometimes publish papers in your field – there would be far too many for one thing, and in any case a lot of them would not be obtainable on-site. The best course of action is to pick out a handful of the most relevant journals – anything between six and twelve in number – and make certain that you check these systematically. One or two of these you may receive regularly as part of a society membership, a few might be subscribed to by your research group or department and some should be available in your library or information unit.

Library services

Making the time to go regularly to a library to check for new issues of journals is not always as easy as it sounds. It is worth asking the library staff if they provide any services which would help you to keep up with new issues. Some industrial and government-based libraries circulate newly received journals amongst staff. Each journal will have a circulation list, generally starting with the most senior person; you can always ask to have your name added to the bottom. Waiting for journals being circulated like this can be a slow business, of course. Many libraries prefer to alert those interested by making up bulletins of the title pages of newly received issues of journals and circulating these instead. Practices differ between libraries, so again the important thing is not to be afraid of asking about such services. Remember it is in a library's interest to make sure its journals are used as much as possible.

Current contents

Browsing through the ten or twelve most central journals in your subject field will help you to pick up a good number of the most relevant papers being published. Inevitably, though, a fair number of interesting ones will be published in journals on the fringes of your subject field. How can you find out about these papers? What can you do about those journals you want to see regularly but which are not taken by your library? One very popular way of finding such papers is to use a group of journals called *Current contents* published by the Institute for Scientific Information and shown in Figure 38. *Current contents* is based on the simple but very effective strategy of reproducing the tables of contents of major journals in a discipline in a form that can be quickly and easily scanned by a reader.

Fig. 38 *Current contents* **series**

As Figure 39 shows, each separate weekly edition includes a list of all the journals whose contents pages have been reproduced that week; the contents page of one of these journals is also shown in the illustration. Many people appear to use *Current contents* simply by scanning through the contents pages, but it is also possible to search for specific authors or subjects using title, word and author indexes.

Table of contents databases
An alternative method of searching is to use the electronic version of *Current contents*, distributed weekly on diskette, and supplied with its own retrieval software. Several other such 'table of contents' databases on diskettes are now available from other producers; a major advantage of such databases is that relevant references can often be directly exported or transferred to personal database software packages, thus avoiding the chore of manually typing in new references.

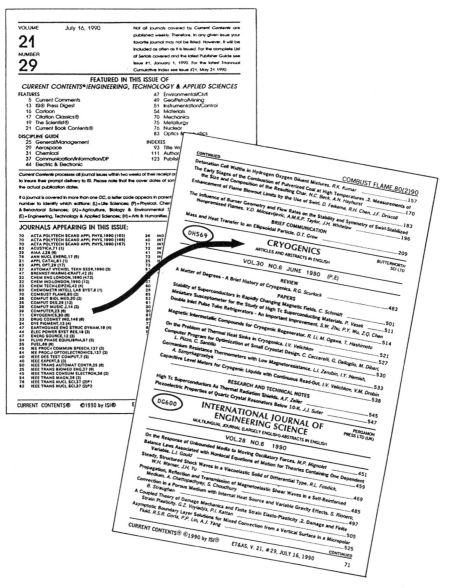

Fig. 39 *Current contents/engineering, technology & applied sciences*

Current awareness bulletins

Scanning the contents pages of journals is a very popular means of keeping up to date with published information, but other equally satisfactory methods can also be used. A variety of newsletters is published with the single purpose of alerting scientists and technologists to new papers, reports and so on in specific, fairly narrow subject fields. Many are actually subsets of abstracts or indexes and their online counterparts. The major expense in the production of databases is in the compiling, indexing and keyboarding of the information; once the database is in existence the more end-products which can be obtained the better.

These newsletters, known as current awareness bulletins, are published under a variety of names; the series produced as a spin-off from *Chemical abstracts*, for example, is called *CA selects*. These types of publication are really aimed at scientists and technologists, not libraries or information units. Sometimes there are neither summaries nor subject indexes as such bulletins are intended for immediate short-term use, not for searching back for information over several years. Leaving out summaries and indexes also helps keep the price down to a level within the reach of a research team or department.

Because there are so many different current awareness bulletins and because new ones appear frequently, it is not possible for us to include a list in this book. The best course of action is to ask at your library or information centre for help in finding out the names of any suitable publications.

SDI services
Current awareness bulletins are a very useful way of keeping up to date provided that one can be found which covers the subject field satisfactorily. If there is nothing which appears suitable, another option is to consider using an SDI service. SDI services – the initials stand for selective dissemination of information – are custom-built services which inform individuals or small groups of people about published information in their own subject field. Sometimes libraries and information centres in industrial or government organizations provide their own clientele with an SDI service by scanning journals, lists of new books and so on manually. The type of SDI services that we are referring to here, though, are computer-based and available to anyone able to pay for them. Specific information about an individual's or group's subject interests (known as its profile) is matched by computer with an appropriate database or bases. Since only new information is wanted, the profile would need to be matched only with the most recent part of the database. This is quite feasible because updates – new additions to databases – are made at regular time intervals. Once the matching process has been carried out, all the references to relevant papers, reports and so on are printed out for scanning by the customer.

Some of the big database producers such as INSPEC and the Institute for Scientific Information sell their own SDI service. Profiles are generally compiled by the database producer's staff, with subscribers completing a form outlining their subject interests in detail.

Drawing up a profile for an SDI search is basically the same procedure as that for a computerized search: the keywords describing the search concepts are linked together using 'AND', 'OR' and 'NOT' logic. It is possible to list the names of any individuals whose work is of interest: any papers they publish which are included in the update to the database will be printed out. It is also possible to exclude papers published in journals which are always seen, such as the ones received through a personal subscription. Subscribers are informed of their references in the form of computer print-out, such as those produced by INSPEC (Figure 40). Once the service is in full operation, a profile needs to be monitored regularly and modified as necessary. If this is not done a profile will gradually become out of date as subject interests shift.

Online hosts, such as DATASTAR, also supply SDI services. The services are essentially the same as those provided by the database producers: profiles are matched with the updates to databases, the relevant references being either downloaded or printed out for scanning by subscribers. The difference is that the computerized matching of profiles with databases is carried out by the online host, not the database producer. Sometimes the plan or strategy of a retrospective search, one dating back over several years, is 'saved' or stored permanently in the host's computer. This can be a very useful way of keeping up with a topic for which most of the prior information has been obtained. It is, though, equally possible to draw up a profile specifically to search for new information.

INSPEC SDI Service P0302   Issue 90003   Page 1

000089   Adaptive neural network in a hybrid optical/electronic architecture using lateral inhibition. P.J.de Groot. R.J.Noll (Dept. of Res., Perkin-Elmer Corp., Danbury. CT. USA).
Appl. Opt. (USA). vol.28. no.18, p.3852-9 (15 Sept. 1989). BLDSC: 1576.25000.
The authors report the optical implementation of a neural network based on a nearest matched filter algorithm and extensive lateral inhibition. Extremely rapid learning is demonstrated in pattern recognition and autonomous control applications. without introducing processing artifacts such as spurious states and ambiguous solutions. The optical implementation is achieved with a reconfigurable. bipolar mask-type crossbar switch based on an inexpensive liquid crystal spatial light modulator. (26 refs.)
Document Type: Journal Paper
Treatment Codes: Theoretical/Mathematical
Classification Codes: A4230S A4280K B4180 B6140C B4150D C1250 C1230
Controlled Index Terms: adaptive filters: liquid crystal devices: matched filters: neural nets: optical information processing: optical modulation: pattern recognition: spatial filters
Free Index Terms: adaptive neural network: extremely rapid learning: reconfigurable bipolar mask type crossbar switch: hybrid optical/electronic architecture: lateral inhibition: optical implementation: nearest matched filter algorithm: pattern recognition: autonomous control applications: liquid crystal spatial light modulator:

Fig. 40 INSPEC SDI service

Cost
Keeping up to date by using a custom-designed SDI service will be more expensive than using a current awareness service where the cost is spread among a number of people. For some services the cost is higher if full abstracts are wanted and the cost is also increased if the output, that is the number of references, exceeds a certain limit. The prices of online SDI services are more variable: the cost will depend on how frequently the profile is matched against the database and whether the output is limited to a maximum number of references.

Value
What benefits do you get from an SDI service for the money? First, provided your profile is kept current, a large percentage of the references you receive will be of direct relevance to you. If you scan a current awareness bulletin or journal contents pages, a fair number of references will be either irrelevant or of fringe interest only. You could, of course, achieve a similar result by scanning through the abstract or index which is the printed equivalent of the database, provided that you check every issue systematically. What an SDI service does is to take much of the effort out of this type of search and – an important point – ensure that it is carried out on a regular basis.

In addition, the references from a computerized SDI search of a database will generally be received well before they appear in the equivalent abstract or index. This is because updating of a database can take place as soon as a sufficient quantity of references is ready for input. The same references will not appear in the equivalent abstract or index until several weeks later because of printing delays. If the publication is being distributed by surface mail across the Atlantic there will be an additional delay of several weeks.

For a new postgraduate student or a lone research worker, the cost of the type of SDI service outlined may be prohibitive. Where a team of people is working on a long-running project, supported by reasonable funding, an SDI service could be very useful. The large number of databases and the different methods of supply make it impracticable to include a list of all possible services here. What we have tried to do is give an impression of what these services are and which organizations supply them, and to indicate their usefulness. If your appetite has been whetted ask your librarian or information officer for help in investigating likely services.

Research in progress
So far in this chapter we have concentrated on how you can keep up with the results of new research and development. However, to get a complete picture of the activity within a speciality you really need some knowledge of what work is currently being carried out: projects which are in progress but about which little or nothing has yet been published.

There are two good reasons why it is useful to have some idea of what is going on:

- It could prevent your duplicating someone else's work.

- More positively it might lead to some assistance, and/or cooperation with your project.

Obviously there are limits to how much you can find out – no one in a commercial organization, a pharmaceutical company for example, is going to reveal the direction of his or her research effort. Within the publicly supported sector – academic institutions and government-owned research establishments – there is more openness about current activities; the following information on discovering relevant projects consequently relates mainly to these areas.

Most people pick up a good percentage of their knowledge of current work through the 'grapevine' system, talking with friends and colleagues and going to seminars, conferences and the like. To supplement this approach, and for anyone without contacts within a subject field, there are registers or listings of current research projects you can use. One major general register of current UK research in academic and government organizations is *Current research in Britain*, compiled annually by the British Library, and available both as an online database and in book format. The printed format consists of four separate volumes covering the physical, biological and social sciences and the humanities. Each is subdivided according to institution and department with name, subject and other supporting indexes, as Figure 41 shows. Another important R & D database in the UK is B.E.S.T. (British Expertise in Science and Technology). The grant-awarding organizations such as the Medical Research Council in the UK also usually either publish a register of current projects or include details in their annual report. There are other published directories, some national in scope, others limited to specific subjects. In the US the Federal Research in Progress (FEDRIP) database covers current government-funded research within the scientific and technological fields. There is one note of warning about registers of current research, however. It is almost impossible to compile a complete list. Inevitably some people do not report their research and some projects may be excluded because of the classified nature of the research. You can never be completely certain that no one else is working in your field, but you can be far more aware of who your major rivals might be.

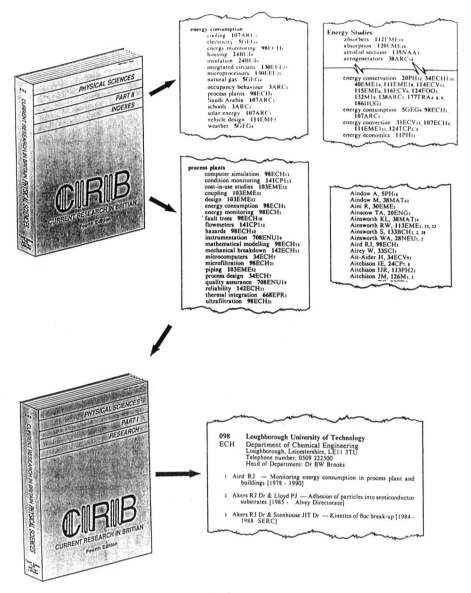

Fig. 41 *Current research in Britain*

98

Summary

- Write to the publishers in your subject area asking to be put on the mailing list for information on new books.

- Pick out the most important journals in your area, and check all new issues of these.

- Supplement this information by using *Current contents*, current awareness bulletins or an SDI service.

- Check registers of research in progress for other relevant projects.

7

Future developments in scientific and technical information

So far we have concentrated on the current sources and methods you can use to find information. In this last chapter we would like to move on from current practices and consider the future. What are the prospects for the structure and communication of scientific and technical information? To what extent will they be affected by the ever-increasing use of information technology?

As we explained at the beginning of the book, scientific and technical information can be viewed from two different, though interrelated, aspects:

- How information is communicated to the scientific and technical community.

- How the individual worker searches out the information needed from the mass of published material.

Both are important to the average scientist or technologist, for even if he or she does not generate new information, he or she will certainly use it and both functions will undoubtedly be further affected by new technology.

Journals

For well over 300 years now, journals have been the major means of communication of new scientific and technical information to the world at large, and any discussion of the future of scientific and technical communication needs to consider their likely role. With the growth of information technology in the 1980s there was much interest in the concept of 'electronic journals'. These were to be journals in which the papers submitted would be stored in a computerized database; it was envisaged that subscribers would log into the database to obtain a copy of a paper as no printed equivalent would be published. Currently the prospects for establishing completely electronic journals along these lines look less

101

promising than they did. Technologically it is quite feasible to set up this type of electronic publication, but the obstacles which would have to be overcome are considerable – the economics, the difficulty of motivating authors to submit papers to journals with no established reputation and the basic lack of convenience of use. Conventional printed journals look set, therefore, to continue for the foreseeable future, but with more and more existing, highly used journals also being published in a parallel electronic version as full-text databases.

Development of online versions of current printed journals in the scientific and technical fields has been hampered until recently by the difficulties of including illustrative information – diagrams, photographs, charts and graphs – which is an essential feature of most scientific and technical papers. In relation to text this type of information requires a very large amount of storage space, and the conventional magnetic discs used by computers do not have sufficient capacity to cope with these requirements. As a result only the text and tables from a paper could be included in the electronic version. Any photographs, diagrams and so on have to be supplied separately, a big disincentive to use. With the introduction of optical discs as computer storage media this situation is beginning to change. Optical discs have a much higher storage capacity, making it feasible to include the diagrams and so on from a paper in the electronic version of the journal.

These types of database on optical disc could be used in one of two ways:

1 Where there would be a high level of use, a library could buy the discs; someone wanting to find a particular paper could search the disc using a microcomputer, display it on a high resolution screen and then print off a permanent copy on a laser printer.

2 Where there would be a low level of use – if, for instance, the database contained papers from journals of fringe interest to the scientists and technologists in an establishment – then a library could obtain the paper from a supply centre which would keep the whole range of discs. In the near future this would entail the supply centre printing off a copy and then despatching it by post. However, such a procedure would really only be an interim solution. The eventual aim of such a system would be to transmit the contents of the paper (both text and illustrations) via a telecommunications network to a workstation at the requestor's organization, either his or her own or one in the library. However, as we mentioned earlier, the computerized versions of papers containing both text and illustrations constitute a relatively large volume of electronic data; the current generation of telecommunications networks do not basically have the capacity to cope with the transmission of such large amounts of data. In the UK a new high capacity network, known as ISDN (Integrated Services Digital Network), is gradually being installed by British Telecom. When this is fully operational

it will be possible to transmit large volumes of data from such databases.

This trend towards an ever-increasing amount of information also being 'published' in a computerized version is not, of course, limited to journals. Many encyclopedias and reference books are already searchable either online on on CD-ROM. Far more are likely to become computerized in the future.

Hypertext

Perhaps the most novel prospect on the horizon is the introduction of hypertext databases. These are not just re-runs of existing printed publications but are an entirely new concept, requiring a computer for both storage and viewing, and distinctive because of the way in which the text and illustrations would be read and studied. In a conventional publication or database the information is read in a linear or sequential manner, that is, the user starts at the beginning of a section of text and reads or scans through it as far as appropriate. A hypertext database, however, could be read in a nonlinear fashion. If, for example, an encyclopedic-type hypertext database on drug production were constructed, a student interested in the manufacture of drugs might start off by studying a description of the economics of drug production. Halfway through this description he or she might come across a reference to patents and, instead of reading to the end of the section, decide to call up on screen the section on the patenting process. Some way through this article, where a mention is made of the regulatory organizations such as the Committee on Safety of Medicines, the student could then decide to skip the rest of the section on patents and display the information on regulatory affairs.

To enable a reader to jump from one piece of information to the next in this way obviously requires relevant, related pieces of information to be linked together within the database; the linkages or associations must be inserted by editors as the database is constructed. The actual process of jumping from one piece of information to the next (provided, of course, that the appropriate links exist) would be virtually instantaneous, requiring only the pointing and clicking of a 'mouse' at the relevant phrase or reference on screen. These hypertext databases are still in their infancy; there are problems to be overcome, not least of which is the expense of construction, but they could prove to be a useful method of storing reference-type information, such as that contained in encyclopedias and handbooks, in computerized form.

Access to information

Looking at the second aspect of scientific and technical information – how the individual worker searches out the information needed from the mass of published material – the trend is towards easier, simpler methods, again using computers. As we mentioned in Chapter 4, the software which is used to search computerized databases is gradually becoming much more 'user-friendly', and is now aimed much less at the professional librarian or information scientist and much more towards the practising scientist or engineer. As more organizations install local area telecommunications networks, linking together all the computing facilities within a site or campus, the possibility of an individual being able to search for and obtain the information needed from a workstation in his or her office becomes one step closer.

In such a scenario a person could, for example, log in and search one of the local library's CD-ROM databases such as MEDLINE. The references obtained could then be checked automatically by the library's computer to see which are available on-site. These could then be transmitted over the network to the individual's workstation. References to papers not in the library could be automatically transmitted via a long-distance network to a supply centre. Here full-text journal databases would be automatically searched by computer to locate copies of the papers, which would then be transmitted back over the network to the individual's workstation. Such a scenario is not yet really feasible either economically or technologically but would certainly be the dream for the twenty-first century.

Index

abstracts 5
  arrangement 33; choice of service
  40–3; information on use
  31–45; list of 41; structure 31,
  32
ASCII format 82
askSam 85
author indexes 39
authority lists 84

B.E.S.T. 97
bibliographies 19–25
Bib/SEARCH 83, 85
Biological abstracts
  subject arrangement 33, 35
Biological abstracts on compact disc 54
BNB see British national bibliography
Bookbank CD-ROM 21
books 4
  British, current titles 20–3; older
  titles 23–5; references 6–7
Books in print PLUS 21
Boolean logic 58, 59, 84
British books
  current titles 20–3; older titles
  23–5
British Expertise in Science and
  Technology *see* B.E.S.T.
British Library 73–4, 76
British national bibliography 23–4
British Standards Institution 4

Cardbox 80–3, 85
CA selects 94
catalogues 15–19

  name 15; title 15
CD-ROM systems 21, 51, 54, 55
  Science citation index 45–8;
  Sigma-Aldrich material safety
  data sheet 70
Chemical abstracts
  index guide 38; Registry Numbers
  38, 61; sample record 31, 32
Citation index 45–8
citations *see* references
classification
  Dewey Decimal 15; *Physics
  abstracts* 33, 34
command-driven systems 61, 62
communication of information
  1–5, 101–3
COMPENDEX*PLUS 54, 60, 66–8
computerized information searching
  51–67
computerized personal literature
  collections 78–80
concepts, computerized searching
  58, 59, 67, 68
conference proceedings 3
  references 7
controlled vocabulary indexes 38
copyright legislation 76–7
costs, online systems 54, 57
cumulated subject indexes 36
Cumulative book index 23, 25
current awareness bulletins 91, 93,
  94
current books 20–3
Current contents 91–3
current information 89–99

current research and development
  96–8
Current research in Britain 97, 98

data communication network 52, 53
databases 5
  choice 60; full-text 69; future
    developments 101–3; list of 41;
    magnetic tape 51; numeric 69,
    70; personal systems, manage-
    ment software 78–86; table
    of contents 92; updating 57
DATASTAR 52, 66, 67
Dewey Decimal Classification 15
DIALOG OnDisc™ MEDLINE 62–6
DIALOG^R 52
Dissertation abstracts international 28
downloading 53, 77, 79

Electrical & electronics abstracts 56
Electronic engineer's reference book 12,
  14
electronic journals 101–2
Elsevier mailing list 90
encyclopedias 12
end-users 55
Engineering index
  sample record 31, 32
ESA-IRS 52
Europhysics letters 45–7

fax 77
Federal research in progress 97
filing rules, library catalogues 17–18
fixed length fields 82
full-text databases 69
future developments 101–4

GEOREF 60
Get-A-Ref™ 85
*Government reports announcements
  and index* 43
guides to published information
  12, 14

handbooks 12
Harvard system 6
hints on searching 43–5
hosts 52–5
  SDI services 95, 96
hypertext 103

IdeaList 85
Index medicus
  CD-ROM database 54; sample
    record 31–2; subject arrange-
    ment 33, 36
Index to scientific reviews 25
Index to theses 25–7
indexes 5
  arrangement 33; author 39;
    choice of service 40–3;
    controlled vocabulary 38;
    cumulated subject 36;
    information on use 31–45;
    list of 41; natural language 38,
    39; patent 40; structure 31–2;
    subject 18
indexing
  manual system 77–8; methods of
    38–9; personal databases
    82–4
INSPEC
  database 56, 57, 60; *Electrical &
    electronics abstracts* 56; *List of
    journals* 42; *Physics abstracts*
    33, 34; SDI services 95
Institute for Scientific Information
  Current contents 91–3; *Index
    to scientific reviews* 26; *Science
    citation index* 43–5
intelligent gateways 60
interlibrary loan 73–4
International Standard Book Number
  17
ISBN *see* International Standard
  Book Number
ISI *see* Institute for Scientific
  Information
ISR *see Index to scientific reviews*

journal catalogues 19, 20
journals
  communication of information
    1–2; future developments
    101–3; references 7; scanning
    91

keeping up to date 89–99
keywords
  computerized searching 60–1;
    manual searching 43, 44;
    personal databases 83

learned societies 2
letter-by-letter, filing rules 17 – 18
library catalogues 15 – 19
lists of published books 19 – 25
literature collections, personal
    organization 77 – 86

magnetic discs 51
mailing lists 90
manual indexing system 77 – 8
manual searching 31 – 45
    choice 55 – 7; keywords 43, 44
McGraw-Hill encyclopedia of science
    & technology 12, 13, 69
Medical subject headings 33, 37
MEDLINE 54, 60
    DIALOG OnDisc™ 62 – 6
menu-driven systems 61
MeSH see Medical subject headings
METADEX 60
Metals abstracts and index 39

New scientist 3
name catalogues 15
natural language indexes 38 – 9
numeric databases 69 – 70
numeric system, references 5

offprints 74, 75
older books 23 – 5
online catalogues 15, 16
online document ordering 76
online systems 52 – 4, 55
    costs 54
optical discs 102
organization, personal literature
    collections 77 – 86

Papyrus 85
patent indexes 40
patents 3
    references 8
people, communication of
    information 1
Periodical title abbreviations 7
personal database management
    software 80 – 6
personal literature collections,
    organization 77 – 86
photocopying copyright legislation
    77

Physics abstracts
    classification system 33, 34
previously known work 11 – 12
Pro-Cite 81, 85
professional institutions 2
proximity operator 66 – 8
publisher's advertising 90, 91
Pure and applied science books 23

record format
    INSPEC database 56 – 7; personal
    databases 81, 84 – 5
records, addition to personal
    database 82
refereeing 2
Reference Manager 80, 85
references 5 – 8
Registry Numbers 38, 61
relevancy 68
reports 3
    references 7
references
    obtaining on interlibrary loan
    73 – 5; recording of information
    44; to previously known work
    11, 12; setting out 6 – 8
reprints *see* offprints
research and development, current
    96 – 8
research in progress 96 – 8
reviews 4 – 5
    tracing 26

scanning journals 91
SCI *see Science citation index*
Science citation index 45 – 8
Scientific American 2
SDI services 94 – 6
searching
    computerized information 51 – 71;
    initial stages 11 – 29; keywords,
    computerized 60, 61; keywords,
    manual 43, 44; manual 31 – 45;
    methods 55 – 8; personal
    databases 84, 85
selective dissemination of
    information *see* SDI
Sigma-Aldrich material safety data
    sheet 70
software, personal database
    management 80 – 6
standards 4
    references 8

starting from scratch 12 – 14
starting paper, *Science citation index*
    45, 46
subject catalogues 19
Subject guide to books in print 21, 23
subject indexes 18, 36, 38, 39

table of contents, databases 92
telecommunications network 52, 53
thesauri 33, 37
theses 3
    references 7; tracing 26 – 8
title, catalogues 15
trade literature 4

truncation 84

*Ulrich's international periodicals
    directory* 40
urgent information, interlibrary loan
    service 76

variable length fields 82
Venn diagrams 58, 59
verbal information 1, 89 – 90

Whitaker's books in print 20 – 2
word-by-word, filing rules 17 – 18
World patents index gazette service 43